Medical Application of Polymer-Based Composites

Medical Application of Polymer-Based Composites

Editor

Haw-Ming Huang

MDPI • Basel • Beijing • Wuhan • Barcelona • Belgrade • Manchester • Tokyo • Cluj • Tianjin

Editor
Haw-Ming Huang
Taipei Medical University
Taiwan

Editorial Office
MDPI
St. Alban-Anlage 66
4052 Basel, Switzerland

This is a reprint of articles from the Special Issue published online in the open access journal *Polymers* (ISSN 2073-4360) (available at: https://www.mdpi.com/journal/polymers/special_issues/ Med_poly_compos).

For citation purposes, cite each article independently as indicated on the article page online and as indicated below:

LastName, A.A.; LastName, B.B.; LastName, C.C. Article Title. *Journal Name* **Year**, *Volume Number*, Page Range.

ISBN 978-3-0365-0260-1 (Hbk)
ISBN 978-3-0365-0261-8 (PDF)

© 2021 by the authors. Articles in this book are Open Access and distributed under the Creative Commons Attribution (CC BY) license, which allows users to download, copy and build upon published articles, as long as the author and publisher are properly credited, which ensures maximum dissemination and a wider impact of our publications.

The book as a whole is distributed by MDPI under the terms and conditions of the Creative Commons license CC BY-NC-ND.

Contents

About the Editor vii

Haw-Ming Huang
Medical Application of Polymer-Based Composites
Reprinted from: *Polymers* **2020**, *12*, 2560, doi:10.3390/polym12112560 1

Minal Thacker, Ching-Li Tseng, Chih-Yen Chang, Subhaini Jakfar, Hsuan Yu Chen and Feng-Huei Lin
Mucoadhesive *Bletilla striata* Polysaccharide-Based Artificial Tears to Relieve Symptoms and Inflammation in Rabbit with Dry Eyes Syndrome
Reprinted from: *Polymers* **2020**, *12*, 1465, doi:10.3390/polym12071465 7

Hsin-Ta Wang, Po-Chien Chou, Ping-Han Wu, Chi-Ming Lee, Kang-Hsin Fan, Wei-Jen Chang, Sheng-Yang Lee and Haw-Ming Huang
Physical and Biological Evaluation of Low-Molecular-Weight Hyaluronic Acid/Fe_3O_4 Nanoparticle for Targeting MCF7 Breast Cancer Cells
Reprinted from: *Polymers* **2020**, *12*, 1094, doi:10.3390/polym12051094 25

Ivan Bessonov, Anastasia Moysenovich, Anastasia Arkhipova, Mariam Ezernitskaya, Yuri Efremov, Vitaliy Solodilov, Peter Timashev, Konstantin Shaytan, Alexander Shtil and Mikhail Moisenovich
The Mechanical Properties, Secondary Structure, and Osteogenic Activity of Photopolymerized Fibroin
Reprinted from: *Polymers* **2020**, *12*, 646, doi:10.3390/polym12030646 41

Yu-Min Huang, Yi-Cheng Lin, Chih-Yu Chen, Yueh-Ying Hsieh, Chen-Kun Liaw, Shu-Wei Huang, Yang-Hwei Tsuang, Chih-Hwa Chen and Feng-Huei Lin
Thermosensitive Chitosan–Gelatin–Glycerol Phosphate Hydrogels as Collagenase Carrier for Tendon–Bone Healing in a Rabbit Model
Reprinted from: *Polymers* **2020**, *12*, 436, doi:10.3390/polym12020436 59

Yu-Jui Fan, Fu-Lun Chen, Jian-Chiun Liou, Yu-Wen Huang, Chun-Han Chen, Zi-Yin Hong, Jia-De Lin and Yu-Cheng Hsiao
Label-Free Multi-Microfluidic Immunoassays with Liquid Crystals on Polydimethylsiloxane Biosensing Chips
Reprinted from: *Polymers* **2020**, *12*, 395, doi:10.3390/polym12020395 75

Eisner Salamanca, Chia Chen Hsu, Wan Ling Yao, Cheuk Sing Choy, Yu Hwa Pan, Nai-Chia Teng and Wei-Jen Chang
Porcine Collagen–Bone Composite Induced Osteoblast Differentiation and Bone Regeneration In Vitro and In Vivo
Reprinted from: *Polymers* **2020**, *12*, 93, doi:10.3390/polym12010093 83

Tsai-Hsueh Leu, Yang Wei, Yi-Shi Hwua, Xiao-Juan Huang, Jung-Tang Huang and Ren-Jei Chung
Fabrication of PLLA/C_3S Composite Membrane for the Prevention of Bone Cement Leakage
Reprinted from: *Polymers* **2019**, *11*, 1971, doi:10.3390/polym11121971 99

Denise Murgia, Giuseppe Angellotti, Fabio D'Agostino and Viviana De Caro
Bioadhesive Matrix Tablets Loaded with Lipophilic Nanoparticles as Vehicles for Drugs for Periodontitis Treatment: Development and Characterization
Reprinted from: *Polymers* **2019**, *11*, 1801, doi:10.3390/polym11111801 117

Sheng-Tung Huang, Nai-Chia Teng, Hsin-Hui Wang, Sung-Chih Hsieh and Jen-Chang Yang
Wasted *Ganoderma tsugae* Derived Chitosans for Smear Layer Removal in Endodontic Treatment
Reprinted from: *Polymers* **2019**, *11*, 1795, doi:10.3390/polym11111795 . . . . **137**

Yun-Liang Chang, Chia-Ying Hsieh, Chao-Yuan Yeh and Feng-Huei Lin
The Development of Gelatin/Hyaluronate Copolymer Mixed with Calcium Sulfate, Hydroxyapatite, and Stromal-Cell-Derived Factor-1 for Bone Regeneration Enhancement
Reprinted from: *Polymers* **2019**, *11*, 1454, doi:10.3390/polym11091454 . . . . **147**

Yu-Chih Huang, Kuen-Yu Huang, Wei-Zhen Lew, Kang-Hsin Fan, Wei-Jen Chang and Haw-Ming Huang
Gamma-Irradiation-Prepared Low Molecular Weight Hyaluronic Acid Promotes Skin Wound Healing
Reprinted from: *Polymers* **2019**, *11*, 1214, doi:10.3390/polym11071214 . . . . **165**

About the Editor

Haw-Ming Huang received his Ph.D. in biomedical sciences from the Taipei Medical University, in 2005. He is currently a professor at the School of Dentistry and Graduate Institute of Biomedical Optomechatronics, Taipei Medical University, Taipei, Taiwan. His main areas of research are those concerning biomedical materials, including applications of biopolymers and low-molecular-weight hyaluronic acid in tissue engineering. He is also particularly interested in non-destroy and non-invasive detection methods in biomaterials.

Editorial

Medical Application of Polymer-Based Composites

Haw-Ming Huang [1,2]

1. School of Dentistry, College of Oral Medicine, Taipei Medical University, Taipei 11031, Taiwan; hhm@tmu.edu.tw; Tel.: +886-291-937-9783
2. Graduate Institute of Biomedical Optomechatronics, College of Biomedical Engineering, Taipei 11031, Taiwan

Received: 24 August 2020; Accepted: 11 October 2020; Published: 31 October 2020

Composites are materials composed of two or more different components. When polymers are combined with other materials, properties of individual polymers such as mechanical strength, surface characteristics, and biocompatibility can be improved, making it feasible to manufacture materials with excellent mechanical properties and biological activity. Polymer-based composites are commonly used in aerospace, automobile, military, and sports applications, and are increasingly being used in biomedicine for tissue engineering, wound dressings, drug release, regenerative medicine, dental resin composites, and surgical operations [1–3]. Research topics in this area cover technical issues of how best to integrate these materials and postintegration material properties.

Developing biological materials using bioderived polymers as a matrix is another medical application of polymer-based composites. Many recent studies have reported improvement of specific material properties by mixing biocompatible nanoparticles into bioabsorbable polymers [4–6]. These nanoparticle/polymer composite biomaterials are showing great promise for use in biomedical applications including tissue engineering [7,8] and wound healing [9,10], and can also be used to fabricate contrast agents for medical imaging in cancer detection or as targeting materials for precise cancer treatment [11]. With this background in mind, this Special Issue entitled Medical Application of Polymer-Based Composites brings together 11 original articles representing recent progress and new developments in the field of biopolymer-based composites.

For tissue engineering researchers working on bone regeneration technology for periodontics, implantology, and maxillofacial surgery, developing new substitute bone grafts that show strong bone growth activity is an ongoing challenge. In this regard, Salamanca et al. developed and tested the performance of collagenated porcine grafts and found that collagenated porcine graft not only induced osteoblast differentiation in vitro, but also exhibited guided bone regeneration in vivo [12]. They demonstrated that collagenated porcine graft has potential to act as an osteoconductive material that can be used for different dental guided bone regeneration procedures.

Poly-L-lactide (PLLA) is a biodegradable polymer widely used in orthopedic devices and dental rehabilitation [13]. Although it exhibits excellent biocompatibility, its insufficient mechanical strength limited its application in load-bearing area. In order to increase the osteoregeneration and accelerate the formation of new bone, bioactive materials have been added to PLLA [14]. For example, to prevent leakage of polymethyl-methacrylate (PMMA) bone cement during kyphoplasty treatment, Leu et al. developed a novel membrane by mixing PLLA and tricalcium silicate as a barrier [15]. They showed that the mechanical and antidegradation properties of this hybrid composite were improved without affecting its cytocompatibility and created a potential antileakage membrane solution for kyphoplasty treatment of osteoporosis related spine fractures.

Soft tissue repair is another important tissue engineering challenge. To develop a collagenase carrier for tendon–bone healing, Huang et al. developed thermosensitive chitosan–gelatin–glycerol phosphate hydrogels. After testing in an animal model, they suggested this novel composite as a potential biomaterial to assist tendon-to-bone healing [16].

Biological polymers are produced naturally in living organisms. They are large molecules composed of many small organic monomers. The use of biological polymers from waste to produce new materials for medical use is a trend for sustainability. Since chitosan and its derivatives have been reported to be a material with high chelating capacity [17], Huang et al. designed an irrigant system using chitosans derived from waste Ganoderma tsugae for smear layer removal in endodontic treatment and concluded that chitosan derived from fungal biomass shows potential as an alternate irrigant for clinical use [18]. Thacker et al.'s study testing physical properties and cell response of artificial tears produced using Bletilla striata polysaccharide is another example of natural plant-based polymers with tissue engineering applications. Their findings indicated that fabricated artificial tears can effectively reduce inflammatory cytokines and reactive oxygen species (ROS) levels in tested cells [19].

Hyaluronic acid (HA) is also a biopolymer found in the various living organisms [20–22] that has shown potential for bone regeneration treatment. The primary physiological characteristics of HA is excellent viscoelasticity and water absorption [23]. Chang and his coworkers developed a gelatin/hyaluronate copolymer composite and mixed it with calcium sulfate, hydroxyapatite, and stromal-cell-derived factor-1 for bone regeneration enhancement [24]. After in-vitro mesenchymal stem cell (MSC) tests and an in-vivo rat model experiment, they found that their Gel-HA/CS/HAP/SDF-1 composite showed an obvious regenerative effect for treating bone defects.

Low-molecular-weight HA (LMWHA) has been reported to exhibit anti-inflammatory effects and accelerated wound repair response [25–28]. To test this, Huang's group prepared LMWHA by gamma-irradiation [29]. After testing LMWHA's chemical and physical properties, the material was combined with carboxymethyl cellulose to form a composite. Their animal experiment showed that wounds in animals covered with the CMC/LMWHA composite decreased in size to almost half that of wounds in animals covered with CMC fabric. Their conclusion suggested that this CMC/LMWHA composite can be an excellent wound dressing material for skin injury.

Nanoparticles have seen recent use in biological applications. Nanomagnetite (Fe_3O_4) is one of these nanoparticles and has attracted attention due to its biocompatibility and functionality [11,14]. In addition, several investigators have manufactured PLLA-Fe_3O_4 composites using electrospinning or injection molding techniques. Since Fe_3O_4 nanoparticles exhibit superparamagnetism, the material has been reported to produce a contrast agent for enhancing the quality of MR tumor images [30]. In addition, HA can target tumor cells through CD44 receptors on tumor cell surfaces [31,32]. Accordingly, in 2020, Wang et al. mixed the natural polymer LMWHA with Fe_3O_4 nanoparticles to target MCF7 breast cancer cells. Using time-of-flight secondary ion mass spectrometry, they proved that LMWHA-Fe_3O_4 nanoparticles have the potential to be used as an injectable agent that can target breast cancer tumors [11].

Drug delivery systems are another application of biopolymer composites. In 2019, Murgia et al. developed a nanostructure lipid carrier to release curcumin [33]. Since curcumin has been reported as a natural polyphenolic compound exhibiting antibacterial, antioxidant, anti-inflammatory, anticancer, and wound healing effects [33–35], this delivery system can be used in the treatment of oral diseases such as oral lesions and periodontitis.

Optically sensitive polymers can also be used as a component to fabricate composites for medical use. Fan et al. developed a new composite using a liquid-crystal-based polydimethylsiloxane substrate. A series of in-vitro tests proved that this design can be used as a detector for multimicrofluidic immunoassays [36]. Bessonov et al. also developed a novel fibroin methacrylamide which can be photocrosslinked into hydrogels and improve their mechanical properties [37]. In addition, their studies showed that their modified polymer increases osteoinductive activity, including increasing cell numbers, rearranging actin cytoskeleton, and improving distribution in focal contacts, indicating that this modified substrate can be used in tissue engineering applications.

References

1. Zafar, M.S.; Najeeb, S.; Khurshid, Z.; Vazirzadeh, M.; Zohaib, S.; Najeeb, B.; Sefat, F. Potential of Electrospun Nanofibers for Biomedical and Dental Applications. *Materials* **2016**, *9*, 73. [CrossRef]
2. Qasim, S.S.B.; Zafar, M.; Najeeb, S.; Khurshid, Z.; Shah, A.H.; Husain, S.; Rehman, I.U. Electrospinning of Chitosan-Based Solutions for Tissue Engineering and Regenerative Medicine. *Int. J. Mol. Sci.* **2018**, *19*, 407. [CrossRef]
3. Zagho, M.M.; Hussein, E.A.; Elzatahry, A. Recent Overviews in Functional Polymer Composites for Biomedical Applications. *Polymers* **2018**, *10*, 739. [CrossRef] [PubMed]
4. Fonseca, A.C.; Serra, A.C.; Coelho, J.F.J. Bioabsorbable polymers in cancer therapy: Latest developments. *EPMA J.* **2015**, *6*, 22. [CrossRef] [PubMed]
5. Wang, H.-T.; Chan, Y.-H.; Feng, S.-W.; Lo, Y.-J.; Teng, N.-C.; Huang, H.-M. Development and biocompatibility tests of electrospun poly-l-lactide nanofibrous membranes incorporating oleic acid-coated Fe_3O_4. *J. Polym. Eng.* **2014**, *34*, 241–245. [CrossRef]
6. Shen, L.-K.; Fan, K.-H.; Wu, T.-L.; Huang, H.-M.; Leung, T.-K.; Chen, C.-J.; Chang, W.-J. Fabrication and magnetic testing of a poly-L-lactide biocomposite incorporating magnetite nanoparticles. *J. Polym. Eng.* **2014**, *34*, 231–235. [CrossRef]
7. Hasan, A.; Morshed, M.; Memic, A.; Hassan, S.; Webster, T.J.; Marei, H.E.-S. Nanoparticles in tissue engineering: Applications, challenges and prospects. *Int. J. Nanomed.* **2018**, *13*, 5637–5655. [CrossRef] [PubMed]
8. Kohane, D.S.; Langer, R. Polymeric Biomaterials in Tissue Engineering. *Pediatr. Res.* **2008**, *63*, 487–491. [CrossRef] [PubMed]
9. Rajendran, N.K.; Kumar, S.S.D.; Houreld, N.N.; Abrahamse, H. A review on nanoparticle based treatment for wound healing. *J. Drug Deliv. Sci. Technol.* **2018**, *44*, 421–430. [CrossRef]
10. Tocco, I.; Zavan, B.; Bassetto, F.; Vindigni, V. Nanotechnology-Based Therapies for Skin Wound Regeneration. *J. Nanomater.* **2012**, *2012*, 1–11. [CrossRef]
11. Wang, H.-T.; Chou, P.-C.; Wu, P.-H.; Lee, C.-M.; Fan, K.-H.; Chang, W.-J.; Lee, S.-Y.; Huang, H.-M. Physical and Biological Evaluation of Low-Molecular-Weight Hyaluronic Acid/Fe3O4 Nanoparticle for Targeting MCF7 Breast Cancer Cells. *Polymers* **2020**, *12*, 1094. [CrossRef] [PubMed]
12. Salamanca, E.; Hsu, C.C.; Yao, W.-L.; Choy, C.-S.; Pan, Y.-H.; Teng, N.-C.; Chang, W.J. Porcine Collagen–Bone Composite Induced Osteoblast Differentiation and Bone Regeneration In Vitro and In Vivo. *Polymers* **2020**, *12*, 93. [CrossRef] [PubMed]
13. Pan, Y.H.; Wang, H.T.; Wu, T.L.; Fan, K.H.; Huang, H.M.; Chang, W.J. Fabrication of Fe_3O_4/PLLA composites for use in bone tissue engineering. *Polym. Compos.* **2017**, *38*, 2881–2888. [CrossRef]
14. Lai, W.-Y.; Feng, S.-W.; Chan, Y.-H.; Chang, W.-J.; Wang, H.-T.; Huang, H.-M. In Vivo Investigation into Effectiveness of Fe_3O_4/PLLA Nanofibers for Bone Tissue Engineering Applications. *Polymers* **2018**, *10*, 804. [CrossRef]
15. Leu, T.-H.; Wei, Y.; Hwua, Y.-S.; Huang, X.-J.; Huang, J.T.; Chung, R.-J. Fabrication of PLLA/C3S Composite Membrane for the Prevention of Bone Cement Leakage. *Polymers* **2019**, *11*, 1971. [CrossRef]
16. Huang, Y.-M.; Lin, Y.-C.; Chen, C.-Y.; Hsieh, Y.-Y.; Liaw, C.-K.; Huang, S.-W.; Tsuang, Y.-H.; Chen, C.-H.; Lin, F.-H. Thermosensitive Chitosan–Gelatin–Glycerol Phosphate Hydrogels as Collagenase Carrier for Tendon–Bone Healing in a Rabbit Model. *Polymers* **2020**, *12*, 436. [CrossRef]
17. Zalloum, H.M.; Mubarak, M.S. Chitosan and chitosan derivatives as chelating agents. In *Natural Polymers, Biopolymers, Biomaterials, and Their Composites, Blends, and IPNs*; Apple Academic Press Inc. Point: Pleasant, NJ, USA, 2013.
18. Huang, S.T.; Teng, N.C.; Wang, H.H.; Hsieh, S.C.; Yang, J.C. Wasted *Ganoderma tsugae* Derived Chitosans for Smear Layer Removal in Endodontic Treatment. *Polymers* **2019**, *11*, 1795. [CrossRef]
19. Thacker, M.; Tseng, C.-L.; Chang, C.-Y.; Jakfar, S.; Chen, H.Y.; Lin, F.-H. Mucoadhesive *Bletilla striata* Polysaccharide-Based Artificial Tears to Relieve Symptoms and Inflammation in Rabbit with Dry Eyes Syndrome. *Polymers* **2020**, *12*, 1465. [CrossRef]

20. Korn, P.; Schulz, M.C.; Hintze, V.; Range, U.; Mai, R.; Eckelt, U.; Schnabelrauch, M.; Moller, S.; Becher, J.; Scharnweber, D.; et al. Chondroitin sulfate and sulfated hyaluronan-containing collagen coatings of titanium implants influence peri-implant bone formation in a minipig model. *J. Biomed. Mater. Res. Part A* **2013**, *102*, 2334–2344. [CrossRef]
21. Correia, C.R.; Moreira-Teixeira, L.S.; Moroni, L.; Reis, R.L.; van Blitterswijk, C.A.; Karperien, M.; Mano, J.F. Chitosan scaffolds containing hyaluronic acid for cartilage tissue engineering. *Tissue Eng. Part C Methods* **2011**, *17*, 717–730. [CrossRef]
22. Dahiya, P.; Kamal, R. Hyaluronic Acid: A Boon in Periodontal Therapy. *N. Am. J. Med. Sci.* **2013**, *5*, 309–315. [CrossRef] [PubMed]
23. Zhang, W.; Mu, H.; Zhang, A.; Cui, G.; Chen, H.; Duan, J.; Wang, S. A decrease in moisture absorption–retention capacity of N-deacetylation of hyaluronic acid. *Glycoconj. J.* **2012**, *30*, 577–583. [CrossRef] [PubMed]
24. Chang, Y.-L.; Hsieh, C.-Y.; Yeh, C.-Y.; Lin, F.-H. The Development of Gelatin/Hyaluronate Copolymer Mixed with Calcium Sulfate, Hydroxyapatite, and Stromal-Cell-Derived Factor-1 for Bone Regeneration Enhancement. *Polymers* **2019**, *11*, 1454. [CrossRef] [PubMed]
25. D'Agostino, A.; Stellavato, A.; Busico, T.; Papa, A.; Tirino, V.; Papaccio, G.; La Gatta, A.; De Rosa, M.; Schiraldi, C. In vitro analysis of the effects on wound healing of high- and low-molecular weight chains of hyaluronan and their hybrid H-HA/L-HA complexes. *BMC Cell Biol.* **2015**, *16*, 19. [CrossRef] [PubMed]
26. Maharjan, A.S.; Pilling, D.; Gomer, R.H. High and Low Molecular Weight Hyaluronic Acid Differentially Regulate Human Fibrocyte Differentiation. *PLoS ONE* **2011**, *6*, e26078. [CrossRef]
27. Rayahin, J.E.; Buhrman, J.S.; Zhang, Y.; Koh, T.J.; Gemeinhart, R.A. High and Low Molecular Weight Hyaluronic Acid Differentially Influence Macrophage Activation. *ACS Biomater. Sci. Eng.* **2015**, *1*, 481–493. [CrossRef] [PubMed]
28. Kavasi, R.-M.; Berdiaki, A.; Spyridaki, I.; Corsini, E.; Tsatsakis, A.M.; Tzanakakis, G.; Nikitovic, D. HA metabolism in skin homeostasis and inflammatory disease. *Food Chem. Toxicol.* **2017**, *101*, 128–138. [CrossRef] [PubMed]
29. Huang, Y.C.; Huang, K.Y.; Lew, W.Z.; Fan, K.H.; Chang, W.J.; Huang, H.M. Gamma-Irradiation-Prepared Low Molecular Weight Hyaluronic Acid Promotes Skin Wound Healing. *Polymers* **2019**, *11*, 1214. [CrossRef]
30. Zhang, H.; Li, J.; Sun, W.; Hu, Y.; Guofu, Z.; Shen, M.; Shi, X. Hyaluronic Acid-Modified Magnetic Iron Oxide Nanoparticles for MR Imaging of Surgically Induced Endometriosis Model in Rats. *PLoS ONE* **2014**, *9*, e94718. [CrossRef]
31. Li, J.; He, Y.; Sun, W.; Luo, Y.; Cai, H.; Pan, Y.; Shen, M.; Xia, J.; Shi, X. Hyaluronic acid-modified hydrothermally synthesized iron oxide nanoparticles for targeted tumor MR imaging. *Biomaterials* **2014**, *35*, 3666–3677. [CrossRef]
32. Zhong, L.; Liu, Y.; Xu, L.; Li, Q.; Zhao, D.; Li, Z.; Zhang, H.; Zhang, H.; Kan, Q.; Sun, J.; et al. Exploring the relationship of hyaluronic acid molecular weight and active targeting efficiency for designing hyaluronic acid-modified nanoparticles. *Asian J. Pharm. Sci.* **2018**, *14*, 521–530. [CrossRef] [PubMed]
33. Murgia, D.; Angellotti, G.; D'Agostino, F.; De Caro, V. Bioadhesive Matrix Tablets Loaded with Lipophilic Nanoparticles as Vehicles for Drugs for Periodontitis Treatment: Development and Characterization. *Polymers* **2019**, *11*, 1801. [CrossRef] [PubMed]
34. Gupta, S.C.; Patchva, S.; Aggarwal, B.B. Therapeutic Roles of Curcumin: Lessons Learned from Clinical Trials. *AAPS J.* **2012**, *15*, 195–218. [CrossRef]
35. Akbik, D.; Ghadiri, M.; Chrzanowski, W.; Rohanizadeh, R. Curcumin as a wound healing agent. *Life Sci.* **2014**, *116*, 1–7. [CrossRef] [PubMed]

36. Fan, Y.-J.; Chen, F.-L.; Liou, J.-C.; Huang, Y.-W.; Chen, C.-H.; Hong, Z.-Y.; Lin, J.-D.; Hsiao, Y.-C. Label-Free Multi-Microfluidic Immunoassays with Liquid Crystals on Polydimethylsiloxane Biosensing Chips. *Polymers* **2020**, *12*, 395. [CrossRef]
37. Bessonov, I.; Moysenovich, A.; Arkhipova, A.; Ezernitskaya, M.; Efremov, Y.; Solodilov, V.I.; Timashev, P.; Shaitan, K.V.; Shtil, A.A.; Moisenovich, A.M. The Mechanical Properties, Secondary Structure, and Osteogenic Activity of Photopolymerized Fibroin. *Polymers* **2020**, *12*, 646. [CrossRef]

Publisher's Note: MDPI stays neutral with regard to jurisdictional claims in published maps and institutional affiliations.

© 2020 by the author. Licensee MDPI, Basel, Switzerland. This article is an open access article distributed under the terms and conditions of the Creative Commons Attribution (CC BY) license (http://creativecommons.org/licenses/by/4.0/).

Article

Mucoadhesive *Bletilla striata* Polysaccharide-Based Artificial Tears to Relieve Symptoms and Inflammation in Rabbit with Dry Eyes Syndrome

Minal Thacker [1,†], Ching-Li Tseng [2,†], Chih-Yen Chang [1], Subhaini Jakfar [1], Hsuan Yu Chen [1] and Feng-Huei Lin [1,3,*]

1. Graduate Institute of Biomedical Engineering, National Taiwan University, No.49, Fanglan Road, Daan District, Taipei 10051, Taiwan; minal.thacker11@gmail.com (M.T.); r05548004@gmail.com (C.-Y.C.); subhaini@yahoo.com (S.J.); hychen83@gmail.com (H.Y.C.)
2. Graduate Institute of Biomedical Materials and Tissue Engineering, Taipei Medical University, No. 250 Wu-Xing Street, Taipei 11031, Taiwan; chingli@tmu.edu.tw
3. Institute of Biomedical Engineering and Nanomedicine, National Health Research Institutes, Miaoli County 35053, Taiwan
* Correspondence: double@ntu.edu.tw; Tel.: +886-928260400
† These authors contributed equally to this work.

Received: 25 April 2020; Accepted: 27 June 2020; Published: 30 June 2020

Abstract: Dry eye syndrome (DES) is a multifactorial disorder of the ocular surface affecting many people all over the world. However, there have been many therapeutic advancements for the treatment of DES, substantial long-term treatment remains a challenge. Natural plant-based polysaccharides have gained much importance in the field of tissue engineering for their excellent biocompatibility and unique physical properties. In this study, polysaccharides from a Chinese ground orchid, *Bletilla striata*, were successfully extracted and incorporated into the artificial tears for DES treatment due to its anti-inflammatory and mucoadhesive properties. The examination for physical properties such as refractive index, pH, viscosity and osmolality of the *Bletilla striata* polysaccharide (BSP) artificial tears fabricated in this study showed that it was in close association with that of the natural human tears. The reactive oxygen species (ROS) level and inflammatory gene expression tested in human corneal epithelium cells (HCECs) indicated that the low BSP concentrations (0.01–0.1% *v/v*) could effectively reduce inflammatory cytokines (TNF, IL8) and ROS levels in HCECs, respectively. Longer retention of the BSP-formulated artificial tears on the ocular surface is due to the mucoadhesive nature of BSP allowing lasting lubrication. Additionally, a rabbit's DES model was created to evaluate the effect of BSP for treating dry eye. Schirmer test results exhibited the effectiveness of 0.1% (*v/v*) BSP-containing artificial tears in enhancing the tear volume in DES rabbits. This work combines the effectiveness of artificial tears and anti-inflammatory herb extract (BSP) to moisturize ocular surface and to relieve the inflammatory condition in DES rabbit, which further shows great potential of BSP in treating ocular surface diseases like DES in clinics in the future.

Keywords: *Bletilla striata* polysaccharide; dry eye syndrome; artificial tear; anti-inflammatory

1. Introduction

Dry eye syndrome (DES) is one of the most prevalent ocular surface diseases in the worldwide affecting 8% to 14% of the population [1,2]. The aging population in Taiwan leads to the increase in the prevalence of DES. DES predominates in females more than male of all age group and the likelihood of developing DES increases with age [3–5]. It also generally appears in patients after cataract surgery and laser-assisted in situ keratomileusis (LASIK) [6,7]. According to International Dry Eye Workshop

(DEWS) II held in 2017, DES was redefined as "Dry eye is a multifactorial disease of the ocular surface characterized by a loss of homeostasis of the tear film, and accompanied by ocular symptoms, in which tear film instability and hyperosmolarity, ocular surface inflammation and damage, and neurosensory abnormalities play etiological roles" [8]. The classification of dry eyes can be categorized into two major subtypes: aqueous deficient dry eye (ADDE), a condition where the lacrimal tear secretion is reduced and hyper evaporative dry eye (EDE), wherein the evaporation of the tear film is uncontrollable with regular lacrimal tear secretion [9,10]. Both these conditions could result in tear hyperosmolarity, which is an important intermediary in causing DES related inflammation. Hyperosmolarity initiates a series of inflammatory events in the epithelial cells which leads to the overexpression of proinflammatory cytokines such as interleukin (IL)-1β, IL-8, IL-6, matrix metalloproteinases (MMPs) and tumor necrosis factor α (TNF-α) on the ocular surface [11–14]. In addition, inflammation and dryness other symptoms of DES include pain, discomfort, visual hinderance, redness, burning sensation and sensitivity towards light [5].

The first line therapy towards the immediate relief from the clinical symptoms of DES is the use artificial tears (AT) [15]. These are the preferred first course of action to ease the symptoms of DES due to their easy accessibility, cost-effectiveness, low side effect and noninvasive nature. AT provide temporary aid by increasing the moisture content and lubricating the ocular surface, in turn reducing the friction between the eye surface and eyelids. However, artificial tears lack in providing substantial long-term relief from DES related inflammation, which is an important underlying issue that needs to be addressed [16,17]. In several cases, anti-inflammatory drugs have been used to treat the underlying DES related inflammation. Anti-inflammatory agents such as topical corticosteroids alleviates the symptoms and inflammation in DES patients, but its use leads to long-term side effects resulting in cataracts and glaucoma [18]. Very few drugs such as lifitegrast and cyclosporine A have been used to reduce the inflammation associated with DES, however their easy accessibility is still an issue in a few countries [1,19]. Therefore, there is a growing need for an alternative agent for the treatment of DES in terms of both inflammation and other symptoms simultaneously.

Due to the safety concerns pertaining to the use of non-steroidal anti-inflammatory drugs (NSAIDs), there is a growing interest to adopt naturally occurring plant-based products. *Bletilla striata* polysaccharide is a plant-based water-soluble polysaccharide isolated from a terrestrial orchid *Bletilla striata* found in the east Asian countries [20]. Polysaccharides have attained limelight due to their various biologic properties, nontoxicity and biodegradable nature [21]. BSP is known to possess anti-inflammatory, antioxidant and antiviral properties [22,23]. Previous study showed that polysaccharides from *Bletilla striata* can successfully alleviate ROS generation and proinflammatory cytokines activation induced by Angiotensin II in human mesangial cells (HMCs), exhibiting antioxidant and anti-inflammatory properties of BSP [24]. Thus, BSP may play a role as an anti-inflammatory and antioxidant agent for DES treatment. It also exhibits excellent moisturizing and lubricating effect due to its mucoadhesive nature [25]. Therefore, we suspect that BSP-based artificial tears have better therapeutic effect towards DES than AT alone due to the mucoadhesive nature, anti-inflammatory and antioxidant properties of BSP, which would increase the retention time of the eye drop on the ocular surface and simultaneously reduce proinflammatory cytokines and ROS generation on the ocular surface.

The aim of this study was to develop a BPS contained artificial tear with anti-inflammatory property for DES treatment. The feasibility and efficacy of BSP-based artificial tear for DES treatment was examined in vitro by coculturing with inflamed HCECs and by the benzalkonium chloride (BAC)-induced rabbit DES model to evaluate its therapeutic effect in vivo.

2. Materials and Methods

2.1. Materials

Bletilla Striata was purchased from Sheng Chang Company (Taoyuan, Taiwan). Dichlorofluorescin diacetate (DCFDA) kit was obtained from Abcam Company (Eugene, OR, USA). keratinocyte-serum free medium (KSFM), bovine pituitary extract (BPE), insulin, trypsin-ethylenediaminetetraacetic acid (EDTA), penicillin/streptomycin and phosphate-buffered saline were obtained from Gibco BRL (Gaithersburg, MD, USA). Epidermal growth factor (EGF) was purchased from Pepro Tech, Inc. (Rocky hill, NJ, USA). Fibronectin, collagen and albumin (FNC) Coating Mix was purchased from Athena Environmental Sciences, Inc. (Baltimore, MD, USA). live and dead kit, Super Script III First-Stand Synthesis System for reverse transcription polymerase chain reaction (RT-PCR), high-capacity cDNA reverse transcription kits, TaqMan Real-Time PCR Master Mixes, Probe, and Trizol were purchased from Thermo Fisher Scientific (Bartlesville, OK, USA). lipopolysaccharide (LPS) was purchased from Sigma-Aldrich (L2880, Rehovot, Israel), Zoletil 50% and 2% Rompun solution were obtained from Virbac Animal Health (Vauvert, Nice, France) and Bayer Korea, Ltd. (Ansan-city, Kyonggi-do, Korea), respectively. Schirmer strips (Tear Touch) were obtained from Madhu Instruments (New Delhi, India). Topical anesthesia solution (0.5% Alcaine1) was obtained from Alcon-Couvreur N.V. (Puurs, Belgium). Fluorescein (FL) paper strips were obtained from HAAG-STREITAG (Köniz/Bern, Switzerland). All the other chemicals were purchased from Sigma-Aldrich.

2.2. Extraction of Bletilla Striata Polysaccharide

BSP was extracted by the methods as previously described [26]. Briefly, dry *Bletilla striata* was homogenized and dispersed in 80 °C double distilled water for 4 h, further filtered to remove impurities. The crude extract was precipitated with 3× 95% (v/v) ethanol overnight. The resultant precipitate was collected by centrifugation and resolved in distilled water. Deproteinization was carried out by adding 1/3 vol. of chloroform/n-butanol (4:1 v/v) and stirred overnight. The above procedure was repeated twice. The final aqueous phase was dialyzed at a molecular weight cutoff of 3000–5000 Da membrane (Orange Scientific, Belgium) and freeze dried to obtain BSP. The structural property analysis of the polysaccharide was performed using Fourier-transform infrared (FTIR) spectrometer (Jasco FT/IR-4200, Tokyo, Japan), ^{13}C and ^{1}H NMR spectra measurements and TGA-thermogravimetric analyzer (Q50, TA instruments, New Castle, DE, USA). The TGA analysis was performed between 25 and 500 °C with a controlled heating rate at 10 °C/min under a nitrogen atmosphere.

2.3. Cell viability of Human Corneal Epithelial Cells (HCECs)

Cell viability assay was performed with live and dead kit (Thermo Fisher Scientific, Bartlesville, OK, USA) according to the manufacturer's protocol. HCECs (1×10^5/well) were seeded in 24-well plates. After culturing overnight, the medium was replaced with media containing different concentration of BSP (0.01–1% v/v). After culturing for one day, the medium containing BSP were discarded and Calcein AM and Ethidium homodimer-1 were added and cell viability was observed using confocal microscope (Olympus IX71, Japan).

2.4. Antioxidant Effect of BSP

The ROS content in cells before/after treated by BSP was tested by DCFDA Cellular ROS detection assay kit (Abcam Company, Eugene, OR, USA) according to the manufacturer's protocol. Briefly, HCECs were seeded in 96-well black plates (2.5×10^4/well) and cultured overnight. The following day, cells were incubated with 200 (μM) H_2O_2 for 45 min and subsequently replaced with DCFDA reagent and treated for another 45 min. DCFDA reagent was replaced with medium containing different concentrations of BSP (0.01–1% v/v) for 1, 2 and 3 hours, respectively. The amount of 2′, 7′-dichlorofluorescein conversion was determined at 485 nm using a microplate reader (Spectramax plus 384 microplate reader, Molecular devices, CA, USA).

2.5. Anti-Inflammatory Effect of BSP by Gene Expression of Inflammatory Cytokines in HCECs

HCECs were seeded in 6-well plate (3×10^5/well). After 24 h incubation, medium was replaced with media containing 500-ng/mL lipopolysaccharide (LPS) for 3 h in order to stimulate inflammation in cells. Unstimulated HCECs were used as control. Further, the medium was replaced with fresh medium containing different concentration of BSP (0.01–1% v/v). The cells were collected after 2 h and total RNA was extracted according to the manufacturer's protocol using TRIzol reagent. The first strand complementary DNA (cDNA) was synthesized from the isolated RNA of different groups stored at −80 °C, with a concentration of 2 µg/µL, respectively, using the high-capacity cDNA Reverse Transcription Kit according to the manufacturer's protocol. Step One Real-Time PCR System (Applied Biosystems, CA, USA) was used to perform real-time PCR using specific primers (IL-6 [Hs00174131m1], IL-8 [Hs 00174103m1], IL-1β [Hs01555413m1], TNF-α [Hs00174128m1] and TaqMan Universal PCR Master Mix (2×) and glyceraldehyde-3-phosphate dehydrogenase [GAPDH; Hs99999905m1]). The relative expression of each target gene was examined by $2^{-\Delta\Delta Ct}$ method.

2.6. Characterization of Artificial Tears with BPS Addition

First, 0.1% BSP was chosen to be incorporated into the basal AT eye drop solution. The composition of 100 mL basal AT solution consisted of 0.45 g NaCl, 0.015 g $CaCl_2$, 0.15 g KCl and 0.45 g Na_2HPO_4. The AT solution was freshly prepared and filtered using 0.22 µm filter. In order to characterize the physical properties of BSP-based AT, refractive index was measured using a refractometer (ABBE T3, ATAGO CO.,LTD, Tokyo, Japan), pH and osmolality were measured using a pH meter (Jenco IE2-6171, Shanghai, China) and an osmometer (Osmotech, Advanced instruments, Norwood, MA USA), while the viscosity was measured using a viscometer/rheometer (HAAKE RheoStress 1, Thermo Scientific, Waltham, MA, USA).

2.7. Ocular Retention Test for BPS Contained AT

In order to examine the retention time of the eye drops on the ocular surface, two mice (C57BL/6 J) were used in the test (two repeated tests with one-day rest interval). Then, 10 µg/mL red fluorescent dye (TAMRA) was added to only AT or AT-containing 0.1% BSP solution to observe and track the distribution of only AT and BSP-based AT on the eye surface. Five microliters each of only AT and BSP-based AT solution containing fluorescent dye was dropped on the respective mice eye, the fluorescent signal and the photographs were recorded using the In vivo imaging system (IVIS-200 imaging chamber, Xenogen, Alameda, CA, USA) at specific time intervals (5, 10, 15, 30, 45 and 60 min).

2.8. Animal Model of DES and Treated by BSP Eye Drops

Eight male New Zealand rabbits (weighing 2.5–3.5 kg) were selected with no indication of ocular inflammation or gross abnormalities. All experimental course of action was in accordance with the institutional animal care and use committee (IACUC) of National Taiwan University College of Medicine and College of Public Health (IACUC approval no. 20170491). The animals were placed in a light-controlled room in standard cages at a temperature of 23 ± 2 °C, with relative humidity of 60% ± 10% and a 12 h light–dark cycle. Animals were fed food and water according to their needs.

All the experimental procedures were conducted under the general anesthesia administered intramuscularly with Zoletil 50 and Rompun (1:2, 1 mL/kg). Then, 20 µL 0.1% benzalkonium chloride (BAC) was instilled three times per day onto the ocular surface of both the eyes of 6 rabbits for four consecutive weeks. The animals were then randomly divided into three groups, with each group containing 4 eyes/2 rabbits and treated with different eye drops: (a) only AT and (b) BSP + AT group containing 0.1% (v/v) BSP in AT solution and (c) 0.1% (v/v) BAC eye drops used as negative control. The fourth group consist of 2 more rabbits (4 eyes) without BAC inducement which were treated as positive controls. These eye drops were instilled onto the ocular surfaces of rabbits for three weeks, thrice a day at regular intervals. After monitoring for three weeks, the appearance of the ocular surface of rabbits was observed via bright field, slit-lamp and fluorescein staining. To further confirm the treatment effect, Schirmer's test and cornea thickness tests were also performed. Thereafter, the rabbits were euthanized, and corneas were excised for histological examination. All the above tests are described below.

2.8.1. Measurement of Tear Production

Schirmer strips were used to measure the aqueous tear secretion [27]. Briefly, anesthesia was administered to the rabbits to keep them motionless. The procedure was conducted in a standard environment by the same person at defined time points. Post topical administration of 0.5% Alcaine®, Schirmer strips were placed by pulling the lower eyelid on the palpebral conjunctival vesica, which is located near the intersection of the middle and outer third of the lower eyelid. The wetted length of the strip was measured (in millimeters) after 5 min. The average length of the wetted study was calculated by testing each eye twice at an interval of 30 min.

2.8.2. Central Corneal Thickness Measurement (CCT)

Using an ultrasonic pachymeter (iPac®pachymeter, Reichert Technologies, Depew, NY, USA) with a hand-held solid probe, cornea thickness was measured. The tip of the probe was held perpendicular to the center of the cornea. Three readings were recorded for each eye and the average was calculated.

2.8.3. Appearance of the Ocular Surface

The ocular surface of the rabbits of each group were observed via slit-lamp microscope after fluorescein staining. For fluorescein staining, 1% fluorescein sodium (2 µL) was dropped onto the ocular surfaces of the rabbits and observed with a hand-held slit-lamp (SL-17, Kowa Company, Ltd., Torrance, CA, USA).

2.8.4. Histological Examination of the Cornea

After euthanizing the rabbits, corneas were fixed for 24 h in 3.7% formaldehyde solution. The fixed corneas were then embedded in paraffin and sectioned. The sections were then stained with hematoxylin and eosin (H&E) and observed under a microscope.

3. Results

3.1. Characterization of BSP

After the extraction from the herbal orchid, the identification results of BPS by FITR, TGA, and NMR is provided here. For FTIR spectra, the characteristic absorption at 874 cm^{-1} corresponds to the β-glucosyl residues. The absorption peak at 809 cm^{-1} suggested that BSP contains mannose while the peaks at 1023 and 1150 cm^{-1} revealed the extant of pyranoses (Figure 1). The present FTIR results are fully in consistence with previous reports [20,26].

For thermal decomposition of BSP, TGA curve was obtained. There are two significant weight losses (Figure 2). The first mass loss is at 100 °C corresponding to the loss of adsorbed water of the sample while the second mass loss is at the onset of 250 °C corresponding to the unique characteristic feature of BSP. These results are also in consistent with the previous reports [28].

The structure of BSP was analyzed by ^{1}H and ^{13}C NMR spectroscopy. The ^{1}H NMR spectra showed two main peaks at δ 4.80 and δ 4.6 ppm (Figure 3a). The chemical shift of carbohydrate indicates the signals for anomeric protons (R). The uncharacteristic peaks found between 3–4 ppm indicates the non-anomeric ring protons. For better elucidation, ^{13}C NMR was performed and the spectra showed two main peaks at δ103.50 and δ100.08 ppm (Figure 3b), revealing the existence of β-glucopyranose and α-mannopyranose repeating units in BSP (R). The chemical shifts found in ^{13}C NMR further confirmed the anomeric carbon. The ^{1}H NMR and ^{13}C NMR chemical shifts were similar to that of an ion exchange chromatography purified BSP [29].

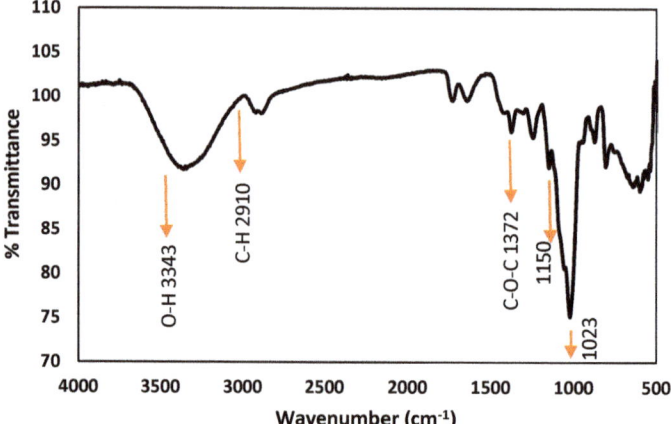

Figure 1. FTIR characterizations of *Bletilla striata* polysaccharide (BSP).

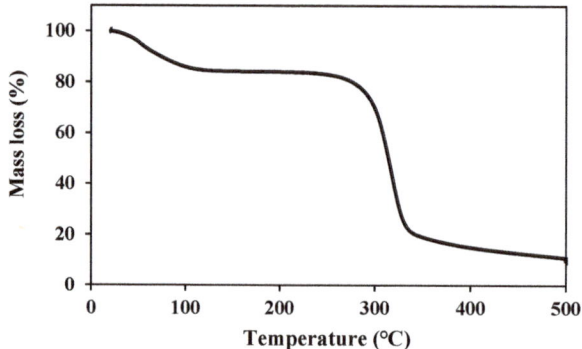

Figure 2. Thermal characterization of BSP (TGA).

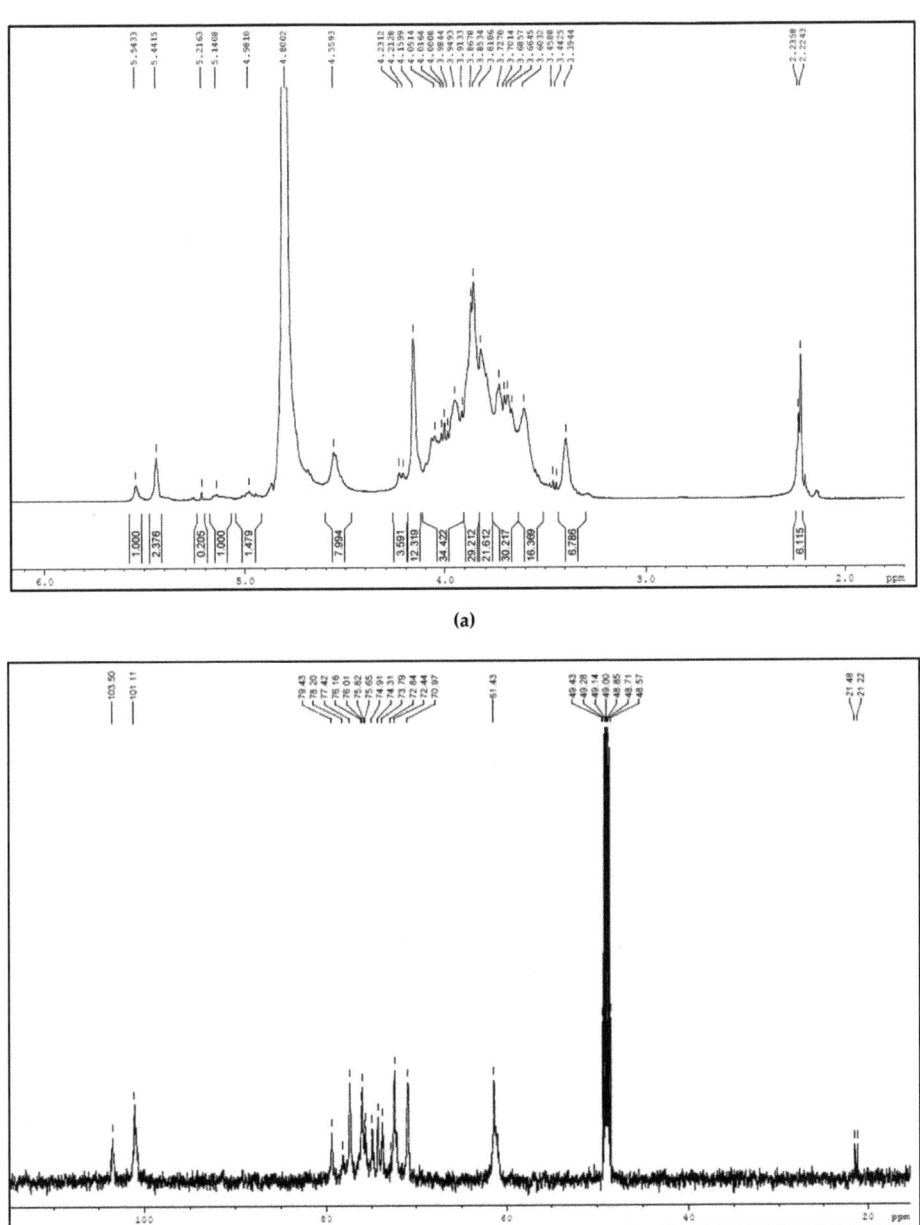

Figure 3. (a) ^{1}H NMR spectrum of BSP; (b) ^{13}C NMR spectrum of BSP.

3.2. Cell Viability of HCEC

The effect of different concentrations of BSP (0.01–1%) on HCECs was demonstrated by live and dead staining (Figure 4). The result shows that HCECs has good cell viability with all the given concentrations of BSP proving the biocompatibility of BPS.

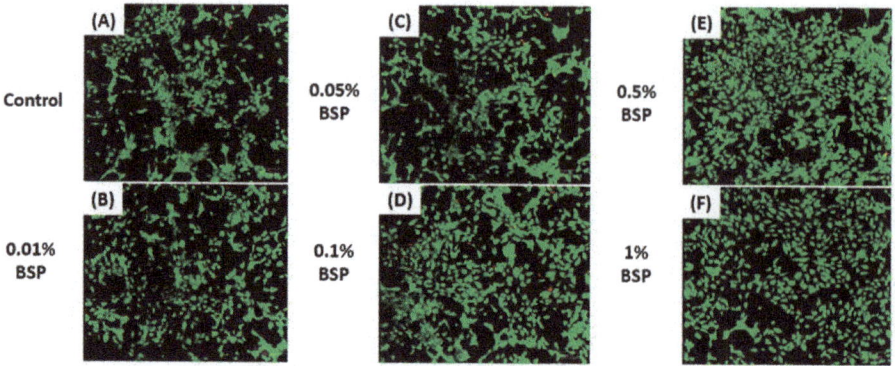

Figure 4. Cell viability of human corneal epithelium cells (HCECs) assessed by live and dead staining (**A**) Control group, HCEC treated with (**B**) 0.01% BSP; (**C**) 0.05% BSP; (**D**) 0.1% BSP; (**E**) 0.5% BSP; (**F**) 1% BSP; scale bar: 200 µm.

3.3. Antioxidant Effect of BSP by DCFDA Cellular ROS Detection Assay

The DCFDA experiment was used to confirm the antioxidant effect of BSP by quantifying the ROS level in HCECs (Figure 5). It was demonstrated that the lower concentrations of BSP (0.01%, 0.05% and 0.1%) can reduce the ROS levels more effectively than the higher concentrations (0.5% and 1%). Therefore, BSP-based AT would be able to reduce the oxidative stress by scavenging ROS on the ocular surface of DES model.

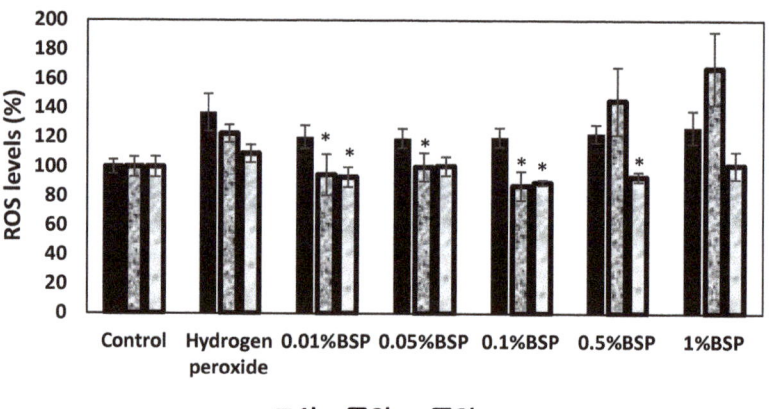

Figure 5. DCFDA cellular R reactive oxygen species (ROS) OS detection assay (n = 6), * $p < 0.05$ compared to hydrogen peroxide group by one-way ANOVA.

3.4. Anti-Inflammatory Effect of BSP by Gene Expression of Inflammatory Cytokines in HCECs

Anti-inflammatory effect of BSP was examined by measuring the levels of the proinflammatory cytokines TNFα, IL8 and IL1β in inflammation stimulated HCECs incubated with LPS and BSP with concentrations 0.01%, 0.05%, 0.1%, 0.5% and 1%. In Figure 6, the expressions of TNFα, IL8 and IL1β were upregulated post LPS inducement in HCECs, which was expected, as LPS treated cells replicates the inflammatory condition associated with DES. However, expressions of TNFα, IL8 and IL1β significantly decreased in cells treated with 0.05% and 0.1% than the cells treated with LPS (* $p < 0.05$). Further, IL8 and IL1β expressions were downregulated in cells treated with 0.01% BSP than in cells treated with LPS (* $p < 0.05$). The overall result shows that the lower concentrations of BSP (0.01%, 0.05% and 0.1%) provide better anti-inflammatory effect by significantly relieving the inflammation in HCECs. Therefore, in view of the above results, 0.1% BSP was chosen to be an optimal concentration for further experiments in the animal model.

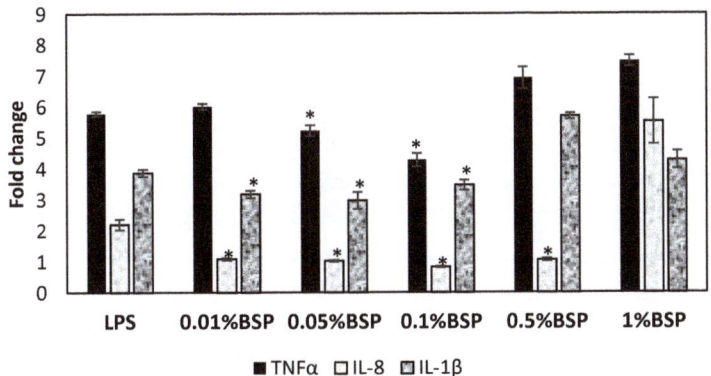

Figure 6. Expressions of inflammatory cytokine (TNF)α, IL8 and IL1β in lipopolysaccharide (LPS)-induced HCEC following different concentrations of BSP treatment. Cells not treated with LPS were used as controls. (n = 3), * $p < 0.05$ compared to LPS group by one-way ANOVA.

3.5. Physical Properties of Artificial Tears

Table 1 shows the summary of the physical properties between human tears [30–32] and BSP artificial tears. The viscosity (4.6–4.7 mPa s) and osmolality (266 ± 0.82 mOsm/kg) of BSP-based AT is within the range of normal human tears, while the refractive index of AT+BSP is 1.3345 ± 0.0015, which is very close to that of real human tears. However, the pH value of artificial tears (7.68 ± 0.01) is slightly out of range when compared to the real human tears, but it is still acceptable.

Table 1. Characteristics of human tears [30–32] and artificial tears.

	Refractive Index	pH Value	Viscosity (mPa s)	Osmolality (mOsm/kg)
Human tears	1.337 (Approx.)	6.5–7.6	1–10	260–340
0.1% BSP+AT	1.3345 ± 0.0015	7.68 ± 0.01	4.6–4.7	266 ± 0.82

3.6. Effect of BSP to Enhance the Retention of AT on the Ocular Surface

A florescent dye (TAMRA) was mixed with AT-containing BSP (BSP+AT) and only AT in order to detect the florescent signal via IVIS to observe its distribution in the mice's eyes. Following 5-min, 10-min, 15-min, 30-min, 45-min and 60-min exposure to the respective eye drops in Figure 7, mice treated with BSP + AT exhibited higher fluorescent intensity than mice treated with only AT. This result indicated that BSP can indeed increase the retention time of the artificial tears and impart lasting lubrication onto the ocular surface, due to the mucoadhesive nature of BSP.

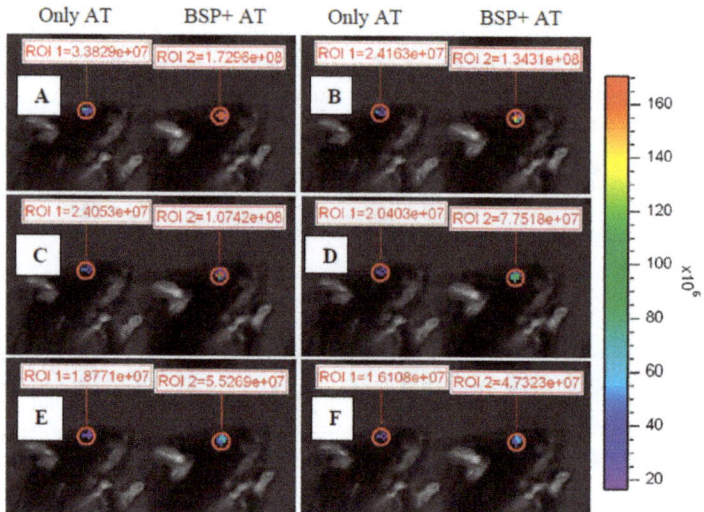

Figure 7. Color photographs showing the ocular retention of the artificial tears mixed with dye (TAMRA) on the ocular surface of mice treated with only artificial tears (AT) and BSP + AT after (**A**) 5-min; (**B**) 10-min; (**C**) 15-min; (**D**) 30-min; (**E**) 45-min and (**F**) 60-min exposure to the artificial tears.

3.7. Appearance of the Ocular Surface

Figure 8 shows the appearance of the ocular surface under a slit-lamp microscope, examined three weeks post the treatment with different eye drops. The ocular surface of the positive control group did not exhibit any fluorescent staining when observed under a slit-lamp microscope. In contrast, BAC eye drops treated group (negative control) exhibited many fluorescent residues on the corneas of the rabbits, indicating inflammation induced by BAC onto the ocular surface. Few staining residues were observed in the corneas treated with only AT eye drops. However, no staining residues were observed in BSP + AT group.

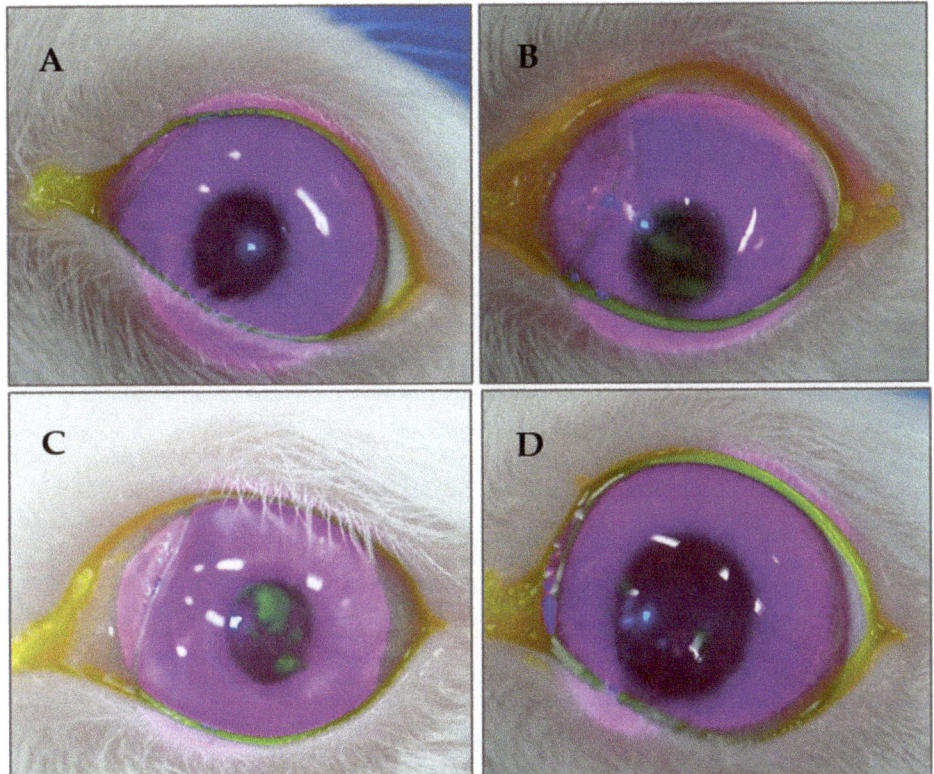

Figure 8. Slit-lamp images of the ocular surface of rabbits after fluorescence staining. (**A**) Control eye; (**B**) eye treated with 0.1% BAC (negative control; dry eye syndrome (DES) eye); (**C**) eye treated with only AT and (**D**) eye treated with AT-containing 0.1% BSP.

3.8. Changes in Tear Production

Tear production is an important indicator for DES evaluation. DES-induced rabbits were examined three weeks after the treatment with different eye drops (Figure 9). Compared to the average wetted length of the positive control group (no DES) (11.33 ± 1.53 mm), wetted average length of BAC group (replicating DES) was 4.7 ± 1.53 mm, indicating decreased tear secretion in the BAC group. The tear production in only AT group had no significant difference (* $p < 0.05$) compared to the BAC group. However, the average wetted length of BSP + AT group was recorded to be 8.3 ± 7.50 mm, indicating increased tear production in BSP treated group than the BAC group.

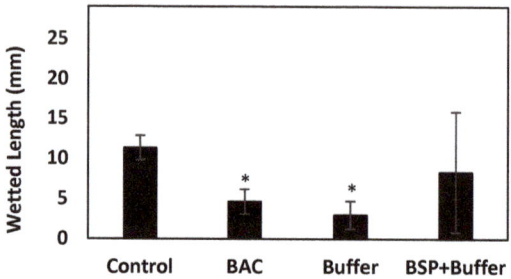

Figure 9. Tear secretion acquired by Schirmer's test (n = 3), * $p < 0.05$ compared to control group by one-way ANOVA.

3.9. Changes in Cornea Thickness

Cornea thickness was measured post three weeks after the treatment with different eye drops. Compared to the positive control group showing average cornea thickness of 349.75 ± 9.49 µm, the average cornea thickness of BAC group increased and was found to be 403.75 ± 6.18 µm, indicating inflammation and swelling of the tissue. The average cornea thickness of only AT group did not show any significant difference compared to the BAC group. However, the average cornea thickness of BSP + AT group was recorded to be 350 ± 8.02 µm, which was similar to the normal cornea thickness (Figure 10).

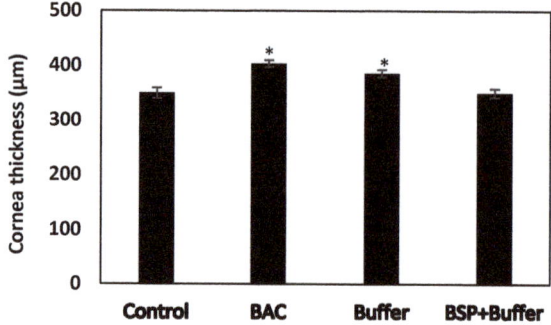

Figure 10. Variation of cornea thickness in rabbits (n = 8), * $p < 0.05$ compared to control group by one-way ANOVA.

3.10. Histological Examination of the Cornea

Examination under a light microscope showed multilayered (three to five layers) epithelial cells with an average thickness of 37.85 ± 3.19 µm and a dense stromal layer containing collagen fibrils in the corneas of the control group (Figure 11A). Defects were seen in rabbits of BAC group (mimicking DES) as it had thinner corneal epithelium consisting of one or two layers with an average thickness of 24.28 ± 2.99 µm (Figure 11B). The only AT group (Figure 11C) and BSP + AT group (Figure 11D) had no significant difference between the corneal epithelium layer, however the stromal layer in BSP + AT treated group was dense than the AT group alone. The tear volume was increased in group treated with BSP + AT than the only AT group indicating therapeutic effects of BSP-based artificial tears on dry eye model.

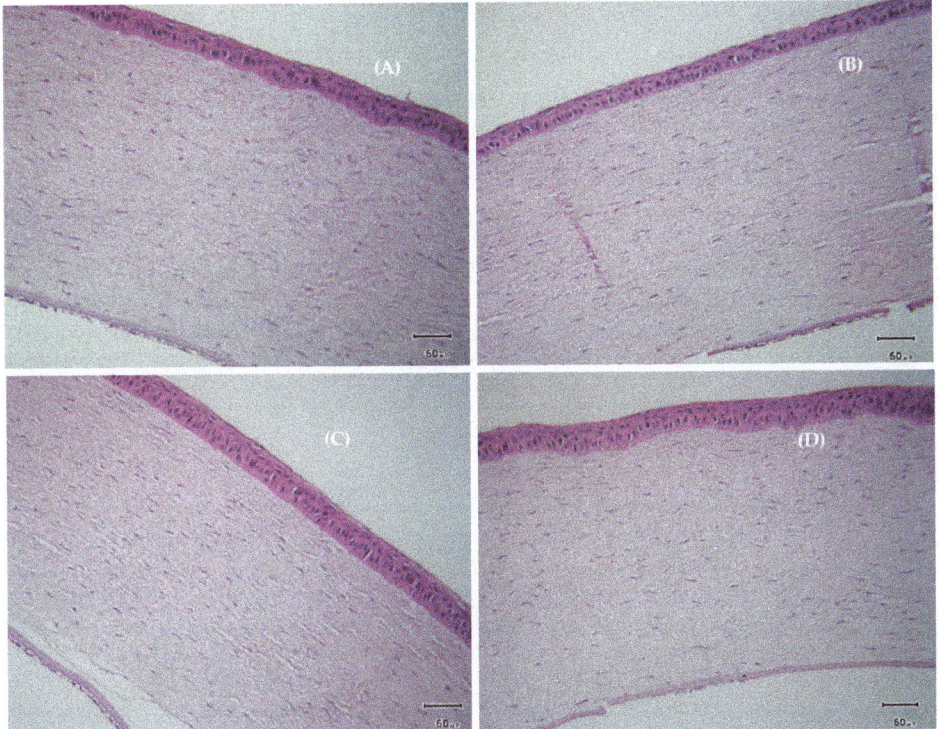

Figure 11. Histological section of cornea with H&E staining acquired from (**A**) control; (**B**) benzalkonium chloride (BAC); (**C**) only AT and (**D**) BSP + AT treated groups, scale bar: 50 μm.

4. Discussion

Artificial tears have gained importance because of their ability to provide immediate ocular comfort in DES. However, most of the available artificial tears fail to address the underlying issue of DES related inflammation. This study confirmed the mucoadhesive and anti-inflammatory effect of BSP-based artificial tears towards effective treatment of DES in rabbit model.

We found that BSP showed a good biocompatibility and was non-toxic to HCECs (Figure 4). This finding was in accordance with the previous study by Wu et al., who showed that BSP had a positive effect on the proliferation of HCECs and demonstrated no cytotoxicity. Dry eye related inflammation is mediated by pro-inflammatory cytokines and inflammatory markers. The most widely studied pro-inflammatory cytokine accompanying dry eye is Interleukin (IL)-1 alongside IL-6, IL-8, IL-1β and TNFα which is also known to play a significant role in DES-related inflammation [33]. These cytokines and inflammatory markers are secreted by lacrimal gland and start the inflammatory reactions causing damage to the epithelial cells on the ocular surface along with the reduction of mucin producing goblet cells causing further damage to the surface [12]. In view of the existing literature, BSP is able to control the levels of pro inflammatory factors in wound healing, thereby suppressing inflammation at the site [34]. Similar results were observed in inflamed HCECs treated with different concentrations of BSP. Lower concentrations of BSP (0.05% *v/v* and 0.1% *v/v*) were able to downregulate the expressions of TNFα, IL8 and IL1β in HCECs (Figure 6), confirming its anti-inflammatory property, which would be beneficial in suppressing DES related inflammation.

Human eye tends to be exposed to oxidative stress due to its high metabolic activity and elevated oxygen tension [35]. Several antioxidants such as uric acid, ascorbic acid, cysteine, lactoferrin present in the tear fluid help protect the ocular surface against certain radicals such as hydroxy and peroxyl radicals [36]. Instability in the tear film due to DES can cause over production of ROS on the ocular surface [37]. Therefore, scavenging ROS is necessary to protect the ocular surface from injury related to oxidization. According to the previous reports, BSP exhibit antioxidant properties and has an ability to scavenge the free radicals, [38]. In this study, DCFDA assay showed that 0.01%, 0.05% and 0.1% BSP can effectively reduce the ROS levels (Figure 5), confirming its antioxidant property, which would help reduce the oxidative stress in DES models.

Mucoadhesion plays a key role in formulating a successful candidate for ocular delivery [39]. BSP is proven to be a mucosa protective agent [25]. The mucoadhesive property of BSP helps to form a protective film on the ocular surface thereby, promoting lubrication and moisturization of the ocular surface, eventually reducing the friction while blinking. In this study, mice were used to demonstrate the retention time and distribution of BSP eye drops on the ocular surface. Mice were used instead of rabbits due to their smaller size which can easily fit in the restricted space of the imaging chamber. Higher fluorescent intensity was observed on the ocular surface of the mice treated with BSP + AT eye drops than ocular surface treated with AT alone (Figure 7) at specific exposure points, indicating BSP's longer retention time on the ocular surface, thereby, providing lasting lubrication and enabling symptomatic relief.

Rabbit animal model is preferred for conducting ophthalmic experiments due to the larger surface area of their exposed ocular surface, making it easily accessible for performing clinical tests like Schirmer's test and fluorescence staining. According to the previous studies, BAC is used to induce DES in animal model [27,40]. However, BAC is also used as a preservative in many ophthalmic eye drops and is shown to aggravate the pre-existing dry eye by causing inflammation and damage to the ocular surface and disturbing the tear film stability [41]. However, inflammation and tear film instability are also a common characteristic of DES, therefore in this study, BAC is used to induce DES in rabbit animal model.

Improvements were observed in recommencing tear volume and cornea thickness in the DES rabbits treated with BSP-based AT. Tear production increased 1.75-fold in rabbits treated with BSP + AT (Figure 9) due to the mucoadhesive nature of BSP that promotes water retention capacity onto the ocular surface. The cornea thickness reduced in the rabbit eyes treated with BSP + AT than the DES-induced eyes (Figure 10) due to the enhanced therapeutic effect of BSP eye drops in reducing swelling and inflammation. Additionally, our study provides the values of refractive index (1.3345 ± 0.0015), pH value (7.68 ± 0.01), viscosity (4.6–4.7 mPa s) and osmolality (266 ± 0.82 mOsm/kg) of BSP-based artificial tears (Table 1), which are all very close to the composition of natural human tears, validating its suitability as a substitute to treat DES. In the present study, H&E examination showed damage to the corneal epithelium (2–3 layers) post BAC inducement (Figure 11B). BAC being a small molecule penetrates through the epithelial layers deep into the stromal layer causing keratocyte damage. The epithelial layers were regained, and stroma was realigned post treatment with AT-containing BSP (Figure 11D).

This study reveals the advantages of using BSP to reduce inflammatory genes expression in inflamed HECEs in vitro, and also to relieve other DES symptoms by recommencing the tear volume, increasing the water retention capacity and promoting corneal recovery in DES model in vivo.

5. Conclusions

The BSP was successfully extracted and formulated into an artificial tear solution for DES treatment. Different concentrations of BSP (0.01–0.1%) cocultured with HCECs showed no toxicity in HCECs. The 0.01%, 0.05% and 0.1% BSP exhibited good antioxidant capacity by lowering the ROS level effectively in stimulated HECEs. While 0.05% and 0.1% BSP demonstrated a good anti-inflammatory effect by significantly downregulating the expression of TNFα, IL8 and IL1β in inflamed HCECs. Therefore, according to the above results, 0.1% BSP was chosen to be the optimal concentration to be mixed with the artificial tear solution for further tests in DES-induced rabbit's model. The BSP artificial tears mimics the natural human tears in terms of similar osmolality, pH, refractive index and viscosity. BSP in AT increased the retention on the ocular surface. The administration of AT-containing BSP effectively promoted corneal recovery and improved tear secretion in DES-induced rabbits to some extent. H&E result shows the recovery of multilayered epithelial cells and stromal layer in corneas of DES rabbits treated with BSP artificial tears. The overall findings of this study suggest that the BSP-based artificial tears not only impart symptomatic relief by providing a good moisturizing and lubricating effect but is also beneficial in providing long term relief from DES-related inflammation which is an underlying issue in the treatment of DES. The present study shows the potential effect of BSP as a promising therapeutic candidate in the treatment of DES. Further clinical examinations are required to evaluate the safety and efficacy of the formulated BSP eye drops for the use in humans to treat DES.

Author Contributions: F.-H.L. conceptualized and administered the research. M.T., C.-L.T. and C.-Y.C. created the methodology and conducted all the experiments. C.-L.T. supervised the DES animal study. H.Y.C. and did the formal analysis of the work. M.T. prepared the original draft of the manuscript. M.T., C.-L.T. and S.J. reviewed and edited the manuscript. All the other authors contributed to data analysis and finalization of the manuscript. All authors have read and agreed to the published version of the manuscript.

Funding: This research received no external funding.

Acknowledgments: The authors would like to thank National Taiwan University College of Medicine and College of Public Health for their support during the animal studies. Assistance of colleagues at the Taipei Medical University are also greatly acknowledged.

Conflicts of Interest: There are no conflicts to declare.

References

1. Abengózar-Vela, A.; Schaumburg, C.S.; Stern, M.E.; Calonge, M.; Enríquez-de-Salamanca, A.; González-García, M.J. Topical Quercetin and Resveratrol Protect the Ocular Surface in Experimental Dry Eye Disease. *Ocul. Immunol. Inflamm.* **2019**. [CrossRef] [PubMed]
2. Moss, S.E.; Klein, R.; Klein, B.E.K. Prevalence of and risk factors for dry eye syndrome. *Arch. Ophthalmol.* **2000**. [CrossRef] [PubMed]
3. Gayton, J.L. Etiology, prevalence, and treatment of dry eye disease. *Clin. Ophthalmol.* **2009**, *3*, 405–412. [CrossRef] [PubMed]
4. Miljanović, B.; Dana, R.; Sullivan, D.A.; Schaumberg, D.A. Impact of Dry Eye Syndrome on Vision-Related Quality of Life. *Am. J. Ophthalmol.* **2007**. [CrossRef]
5. Drew, V.J.; Tseng, C.L.; Seghatchian, J.; Burnouf, T. Reflections on dry eye syndrome treatment: Therapeutic role of blood products. *Front. Med.* **2018**. [CrossRef]
6. Sutu, C.; Fukuoka, H.; Afshari, N.A. Mechanisms and management of dry eye in cataract surgery patients. *Curr. Opin. Ophthalmol.* **2016**, *27*, 24–30. [CrossRef]
7. Levitt, A.E.; Galor, A.; Weiss, J.S.; Felix, E.R.; Martin, E.R.; Patin, D.J.; Sarantopoulos, K.D.; Levitt, R.C. Chronic dry eye symptoms after LASIK: Parallels and lessons to be learned from other persistent post-operative pain disorders. *Mol. Pain* 2015. [CrossRef]
8. Craig, J.P.; Nichols, K.K.; Akpek, E.K.; Caffery, B.; Dua, H.S.; Joo, C.K.; Liu, Z.; Nelson, J.D.; Nichols, J.J.; Tsubota, K.; et al. TFOS DEWS II Definition and Classification Report. *Ocul. Surf.* **2017**, *15*, 276–283. [CrossRef]
9. Villatoro, A.J.; Fernández, V.; Claros, S.; Alcoholado, C.; Cifuentes, M.; Merayo-Lloves, J.; Andrades, J.A.; Becerra, J. Regenerative therapies in dry eye disease: From growth factors to cell therapy. *Int. J. Mol. Sci.* **2017**, *18*, 2264. [CrossRef] [PubMed]
10. Messmer, E.M. Pathophysiology, diagnosis and treatment of dry eye. *Dtsch. Arztebl. Int.* **2015**. [CrossRef]
11. Wilson, S.E.; Perry, H.D. Long-term Resolution of Chronic Dry Eye Symptoms and Signs after Topical Cyclosporine Treatment. *Ophthalmology* **2007**. [CrossRef]
12. Hessen, M.; Akpek, E.K. Dry eye: An inflammatory ocular disease. *J. Ophthalmic Vis. Res.* **2014**, *9*, 240–250.
13. Clayton, J.A. Dry eye. *N. Engl. J. Med.* **2018**. [CrossRef] [PubMed]

14. Tseng, C.L.; Hung, Y.J.; Chen, Z.Y.; Fang, H.W.; Chen, K.H. Synergistic effect of artificial tears containing epigallocatechin gallate and hyaluronic acid for the treatment of rabbits with dry eye syndrome. *PLoS ONE* **2016**. [CrossRef] [PubMed]
15. Baudouin, C.; Galarreta, D.J.; Mrukwa-Kominek, E.; Böhringer, D.; Maurino, V.; Guillon, M.; Rossi, G.C.M.; Van der Meulen, I.J.; Ogundele, A.; Labetoulle, M. Clinical evaluation of an oil-based lubricant eyedrop in dry eye patients with lipid deficiency. *Eur. J. Ophthalmol.* **2017**. [CrossRef] [PubMed]
16. Yagci, A.; Gurdal, C. The role and treatment of inflammation in dry eye disease. *Int. Ophthalmol.* **2014**, *34*, 1291–1301. [CrossRef]
17. Mateo Orobia, A.J.; Saa, J.; Lorenzo, A.O.; Herreras, J.M. Combination of hyaluronic acid, carmellose, and osmoprotectants for the treatment of dry eye disease. *Clin. Ophthalmol.* **2018**, *12*, 456–461. [CrossRef]
18. Kersey, J.P.; Broadway, D.C. Corticosteroid-induced glaucoma: A review of the literature. *Eye* **2006**, *20*, 407–416. [CrossRef]
19. Kunert, K.S.; Tisdale, A.S.; Stern, M.E.; Smith, J.A.; Gipson, I.K. Analysis of topical cyclosporine treatment of patients with dry eye syndrome: Effect on conjunctival lymphocytes. *Arch. Ophthalmol.* **2000**. [CrossRef]
20. Wang, Y.; Liu, D.; Chen, S.; Wang, Y.; Jiang, H.; Yin, H. A new glucomannan from Bletilla striata: Structural and anti-fibrosis effects. *Fitoterapia* **2014**. [CrossRef]
21. Rjeibi, I.; Hentati, F.; Feriani, A.; Hfaiedh, N.; Delattre, C.; Michaud, P.; Pierre, G. Novel antioxidant, anti-α-amylase, anti-inflammatory and antinociceptivewater-soluble polysaccharides from the aerial part of nitraria retusa. *Foods* **2020**, *9*, 28. [CrossRef]
22. Luo, L.; Zhou, Z.; Xue, J.; Wang, Y.; Zhang, J.; Cai, X.; Liu, Y.; Yang, F. Bletilla striata polysaccharide has a protective effect on intestinal epithelial barrier disruption in TAA-induced cirrhotic rats. *Exp. Ther. Med.* **2018**. [CrossRef] [PubMed]
23. Ji, X.; Yin, M.; Nie, H.; Liu, Y. Review Article A Review of Isolation, Chemical Properties, and Bioactivities of Polysaccharides from Bletilla striata. *BioMed Res. Int.* **2020**, *2020*. [CrossRef]
24. Yue, L.; Wang, W.; Wang, Y.; Du, T.; Shen, W.; Tang, H.; Wang, Y.; Yin, H. Bletilla striata polysaccharide inhibits angiotensin II-induced ROS and inflammation via NOX4 and TLR2 pathways. *Int. J. Biol. Macromol.* **2016**. [CrossRef] [PubMed]
25. Wang, L.; Wu, Y.; Li, J.; Qiao, H.; Di, L. Rheological and mucoadhesive properties of polysaccharide from Bletilla striata with potential use in pharmaceutics as bio-adhesive excipient. *Int. J. Biol. Macromol.* **2018**. [CrossRef]
26. Wang, C.; Sun, J.; Luo, Y.; Xue, W.; Diao, H.; Dong, L.; Chen, J.; Zhang, J. A polysaccharide isolated from the medicinal herb Bletilla striata induces endothelial cells proliferation and vascular endothelial growth factor expression in vitro. *Biotechnol. Lett.* **2006**. [CrossRef]
27. Xiong, C.; Chen, D.; Liu, J.; Liu, B.; Li, N.; Zhou, Y.; Liang, X.; Ping, M.; Ye, C.; Ge, J.; et al. A rabbit dry eye model induced by topical medication of a preservative benzalkonium chloride. *Investig. Ophthalmol. Vis. Sci.* **2008**. [CrossRef]
28. Kong, L.; Yu, L.; Feng, T.; Yin, X.; Liu, T.; Dong, L. Physicochemical characterization of the polysaccharide from Bletilla striata: Effect of drying method. *Carbohydr. Polym.* **2015**. [CrossRef]
29. Agrawal, P.K. NMR Spectroscopy in the structural elucidation of oligosaccharides and glycosides. *Phytochemistry* **1992**. [CrossRef]
30. Abelson, M.B.; Udell, I.J.; Weston, J.H. Normal human tear ph by direct measurement. *Arch. Ophthalmol.* **1981**. [CrossRef]
31. Benelli, U.; Nardi, M.; Posarelli, C.; Albert, T.G. Tear osmolarity measurement using the TearLab™ Osmolarity System in the assessment of dry eye treatment effectiveness. *Contact Lens Anterior Eye* **2010**. [CrossRef] [PubMed]
32. Tiffany, J.M. The viscosity of human tears. *Int. Ophthalmol.* **1991**. [CrossRef] [PubMed]
33. Solomon, A.; Dursun, D.; Liu, Z.; Xie, Y.; Macri, A.; Pflugfelder, S.C. Pro- and anti-inflammatory forms of interleukin-1 in the tear fluid and conjunctiva of patients with dry-eye disease. *Investig. Ophthalmol. Vis. Sci.* **2001**, *42*, 2283–2292.
34. Xu, D.; Pan, Y.; Chen, J. Chemical constituents, pharmacologic properties, and clinical applications of bletilla striata. *Front. Pharmacol.* **2019**. [CrossRef] [PubMed]
35. Chen, Y.; Mehta, G.; Vasiliou, V. Antioxidant defenses in the ocular surface. *Ocul. Surf.* **2009**. [CrossRef]

36. Ohashi, Y.; Dogru, M.; Tsubota, K. Laboratory findings in tear fluid analysis. *Clin. Chim. Acta* **2006**, *369*, 17–28. [CrossRef]
37. Dogru, M.; Kojima, T.; Simsek, C.; Tsubotav, K. Potential role of oxidative stress in ocular surface inflammation and dry eye disease. *Investig. Ophthalmol. Vis. Sci.* **2018**. [CrossRef]
38. Qu, Y.; Li, C.; Zhang, C.; Zeng, R.; Fu, C. Optimization of infrared-assisted extraction of Bletilla striata polysaccharides based on response surface methodology and their antioxidant activities. *Carbohydr. Polym.* **2016**. [CrossRef]
39. Graça, A.; Gonçalves, L.M.; Raposo, S.; Ribeiro, H.M.; Marto, J. Useful in vitro techniques to evaluate the mucoadhesive properties of hyaluronic acid-based ocular delivery systems. *Pharmaceutics* **2018**, *10*, 110. [CrossRef]
40. Lin, Z.; Liu, X.; Zhou, T.; Wang, Y.; Bai, L.; He, H.; Liu, Z. A mouse dry eye model induced by topical administration of benzalkonium chloride. *Mol. Vis.* **2011**, *17*, 257–264.
41. Wilson, W.S.; Duncan, A.J.; Jay, J.L. Effect of benzalkonium chloride on the stability of the precorneal tear film in rabbit and man. *Br. J. Ophthalmol.* **1975**. [CrossRef] [PubMed]

© 2020 by the authors. Licensee MDPI, Basel, Switzerland. This article is an open access article distributed under the terms and conditions of the Creative Commons Attribution (CC BY) license (http://creativecommons.org/licenses/by/4.0/).

Article

Physical and Biological Evaluation of Low-Molecular-Weight Hyaluronic Acid/Fe$_3$O$_4$ Nanoparticle for Targeting MCF7 Breast Cancer Cells

Hsin-Ta Wang [1,†], Po-Chien Chou [1,2,†], Ping-Han Wu [3], Chi-Ming Lee [4], Kang-Hsin Fan [5], Wei-Jen Chang [2], Sheng-Yang Lee [2] and Haw-Ming Huang [2,6,*]

1. School of Organic and Polymeric, National Taipei University of Technology, Taipei 10608, Taiwan; htwang@mail.ntut.edu.tw (H.-T.W.); jack19186@hotmail.com (P.-C.C.)
2. School of Dentistry, College of Oral Medicine, Taipei Medical University, Taipei 11031, Taiwan; m8404006@tmu.edu.tw (W.-J.C.); seanlee@tmu.edu.tw (S.-Y.L.)
3. Graduate Institute of Biomedical Materials and Tissue Engineering, College of Biomedical Engineering, Taipei Medical University, Taipei 11031, Taiwan; gahwclbjwph@hotmail.com
4. Core Facility Center, Office of Research and Development, Taipei Medical Universitry, Taipei 11031, Taiwan; jaunslee@tmu.edu.tw
5. Dental Department, En Chu Kong Hospital, New Taipei City 23742, Taiwan; dentaddie@yahoo.com.tw
6. Graduate Institute of Biomedical Optomechatronics, College of Biomedical Engineering, Taipei Medical University, Taipei 11031, Taiwan
* Correspondence: hhm@tmu.edu.tw; Tel.: +886-291-937-9783
† These authors contributed equally to this work.

Received: 25 March 2020; Accepted: 8 May 2020; Published: 11 May 2020

Abstract: Low-molecular-weight hyaluronic acid (LMWHA) was integrated with superparamagnetic Fe$_3$O$_4$ nanoparticles (Fe$_3$O$_4$ NPs). The size distribution, zeta potential, viscosity, thermogravimetric and paramagnetic properties of the LMWHA-Fe$_3$O$_4$ NPs were systematically examined. For cellular experiments, MCF7 breast cancer cell line was carried out. In addition, the cell targeting ability and characteristics of the LMWHA-Fe$_3$O$_4$ NPs for MCF7 breast cancer cells were analyzed using the thiocyanate method and time-of-flight secondary ion mass spectrometry (TOF-SIMS). The experimental results showed that the LMWHA-Fe$_3$O$_4$ NPs were not only easily injectable due to their low viscosity, but also exhibited a significant superparamagnetic property. Furthermore, the in vitro assay results showed that the NPs had negligible cytotoxicity and exhibited a good cancer cell targeting ability. Overall, the results therefore suggest that the LMWHA-Fe$_3$O$_4$ NPs have considerable potential as an injectable agent for enhanced magnetic resonance imaging (MRI) and/or hyperthermia treatment in breast cancer therapy.

Keywords: low molecular weight; hyaluronic acid; TOF-SIMS; nanoparticles; iron oxide

1. Introduction

In clinical contexts, the sensitivity and diagnosis accuracy of magnetic resonance (MR) images are commonly enhanced by injecting a contrast agent into the body prior to scanning [1]. One of the most commonly used agents is gadolinium (Gd); a ductile rare-earth element with ferromagnetic and paramagnetic properties below and above a temperature of 20 °C, respectively [2]. However, with their good biocompatibility, high sensitivity and ease of obtention [3], magnetic nanoparticles (NPs) have attracted increasing interest for a wide range of biomedical applications nowadays, including hyperthermia therapy, drug delivery, biochemical sensing, and tissue repair [4]. Among the various magnetic nanoparticles available, magnetite (Fe$_3$O$_4$) exhibits excellent superparamagnetic properties, and is thus increasingly used in place of Gd as a contrast agent for the MR imaging of tumors [5–8].

Hyaluronic acid (HA) is a biopolymer glycosaminoglycan, which is located mainly in the connective tissue, epidermal tissue, neural tissue and joint tissues in biological organisms [9], and serves as the main component of the extracellular matrix [10]. In medical applications, HA is commonly used in the treatment of osteoarthritis and skin wounds. Furthermore, since it readily binds to the CD44 receptors on tumor cells [10,11], HA is also used as a tumor targeting material [12,13]. HA retains excellent viscoelasticity following water absorption, and is thus useful in retaining skin moisture and treating osteoarthritis [14]. However, with a molecular weight of around 2000 kDa, nature HA has a high viscosity and is hence not easily injected into the human body through the blood vessels. Accordingly, several groups have investigated the physiological functions of low molecular weight hyaluronic acid (LMWHA) (80–800 kDa) [15,16]. It was reported that LMWHA exhibits a potentially beneficial effect on wound healing immune response and angiogenesis [17].

Zhang et al. (2014) used HA in combination with magnetic nano-particles as a contrast agent for enhancing the quality of MR tumor images through targeting with the CD44 receptors on the tumor cell surface [18]. Li et al. used polyethyleneimine-stabilized Fe_3O_4 NPs to integrate LMWHA with two different molecular weights (6 and 31 kDa) and showed that both LMWHAs served as effective probes for the MR imaging of cancer cells with overexpressed CD44 receptors [19]. Zhong et al. (2019) used LMWHA with molecular weights of 7, 63 and 102 kDa to fabricate LMWHA-NPs for drug release. The results indicated that the HA-NPs targeted the CD44 receptors in a molecular weight dependent manner [20].

However, while the aforementioned studies evaluated the targeting ability and performance of LMWHA-NPs using in vitro cellular and in vivo animal tests, the physical properties of the LMWHA-NPs, and in particular, the dynamic viscoelastic properties, were not discussed. Consequently, the synthesis and characterization of LMWHA with a specific molecular weight and a low viscosity for use as an injectable contrast agent for MR imaging remains an important concern. Current techniques for manufacturing LMWHA use two main methods to destroy the main bonds of HMWHA, namely physical methods (e.g., ultrasonic, ozone, electron beam, gamma rays, and heat treatment) and chemical methods (e.g., enzymatic and acid degradation) [21–23]. Among these methods, γ-ray irradiation dramatically decreases the dynamic viscosity of the resulting LMWHA under an applied shear rate, and exhibits Newtonian liquid behavior [17,18]. As a result, the feasibility of the LMWHA as an injectable tumor targeting agent is significantly improved. With this regard, the present study also uses a γ-ray technique to fabricate LMWHA. The superparamagnetic Fe_3O_4 nanoparticles (Fe_3O_4 NPs) was then integrated to LMWHA and the physical and biological characteristics of the resulting LMWHA-Fe_3O_4 NPs were systematically examined by X-ray diffraction (XRD), electrophoretic light scattering, viscosity measurements, and superconducting quantum interference device. Finally, the targeting performance of the LMWHA-Fe_3O_4 NPs for MCF7 breast cancer cells was analyzed using the thiocyanate method and time-of-flight secondary ion mass spectrometry (TOF-SIMS).

2. Materials and Methods

2.1. Materials

$FeCl_2·4H_2O$ was purchased from Avantor Performance Materials, Inc. (Allentown, PA, USA). $FeCl_3·6H_2O$, toluene, ammonia solution hydrochloric acid, ammonium persulfate, potassium thiocyanate and oleic acid were purchased from Nacalai Tesque (Kyoto, Japan). Hyaluronic acid (molecular weight 3000 kDa) was purchased from Cheng-Yi Chemical Industry Co. Ltd. (Taipei, Taiwan). All other analytical grade reagents and solvents used were purchased from J.T. Baker (Phillipsburg, NJ, USA). αMEM, L-glutamine, fetal bovine serum, and penicillin-streptomycin were obtained from Gibco (Grand Island, NY, USA). Sodium nitrate was purchased from Merck (KGaA, Darmstadt, Germany). Finally, dimethyl sulfoxide, Triton X-100 and 3-(4,5-Dimethylthiazol-2-yl)-2,5-diphenyltetrazolium bromide (MTT) were obtained from Sigma-Aldrich (St. Louis, MO, USA).

2.2. Preparation of Fe_3O_4 Nanoparticle

Oleic acid-coated Fe_3O_4 nanoparticles were fabricated using the co-precipitation method described in [24]. Briefly, $FeCl_2 \cdot 4H_2O$ and $FeCl_3 \cdot 6H_2O$ were mixed in a ratio of 1:3 and dissolved in degassed distilled water (DD water). After a stirring process, the sample was heated to 85 °C and NH_4OH was added to achieve a final $FeCl_2 \cdot 4H_2O$, $FeCl_3 \cdot 6H_2O$, NH_4OH, water ratio of 1:2.5:2.5:60. As the sample cooled to room temperature, Fe_3O_4 NPs were formed in accordance with the following reaction:

$$Fe^{2+} + 2Fe^{3+} + 8OH^- \rightarrow Fe_3O_4 + 4H_2O$$

Oleic acid was added to the prepared NPs at 85 °C under stirring for 30 min. A strong magnet was then used to separate the oleic-coated Fe_3O_4 NPs from the solution. The collected NPs were washed in DD water and then dried in an oven at 40 °C for 24 h. The morphology of the coated Fe_3O_4 NPs were observed by a transmission electron microscope (TEM, H-600, Hitachi, Ltd., Tokyo, Japan). In addition, the particle size distribution and zeta potential of the NPs were determined using an electrophoretic light scattering device (NanoBrook 90Plus Zeta, Brookhaven Instruments, New York, NY, USA). The measurement results were obtained for samples with a concentration of 0.25 mg/mL and the measurement process was repeated five times for each sample to ensure the reliability of the results. Finally, the structure and crystalline properties of the fabricated Fe_3O_4 nanoparticles were examined using an X-ray diffractometer (D/MAX 2000 PC, Rigaku Co., Tokyo, Japan) for incidence angles in the range of $2\theta = 20°$ to $90°$.

2.3. Preparation of LMWHA-Fe_3O_4 NPs

In accordance with the method described in [17], LMWHA was produced by irradiating the purchased HMWHA with a cobalt-60 irradiator (Point Source, AECL, IR-79, Nordion, Ottawa, ON, Canada). The irradiating condition was set at 22 °C for 20 h with a dose rate of 1 kGy/h. The molecular weight of the produced LMWHA was measured by gel permeation chromatography (GPC). Briefly, LMWHA was added to 0.1 M NaCl to form a 10 mg/mL LMWHA solution. 200 µL of the LMWHA solution was injected into a separation device (Series 200, Perkin Elmer, Waltham, MA, USA) equipped with a chromatography column (SB-806M HQ, Shodex, Kanagawa, Japan) at 25 °C. The flow rate of the mobile phase (0.1 M HPLC grade sodium nitrate) in the column was set to 0.5 mL/min. The GPC signals were collected using a refractive index (RI) detector (Series 200, Perkin Elmer, Waltham, MA, USA) and the calibration curve was obtained using a standard kit (Pullulan ReadyCal Kits, PSS Polymer Standards Service, Mainz, Germany). The molecular weight of the LMWHA sample was then determined using commercial ChromManager 5.8 software (ABDC WorkShop, Taichung, Taiwan).

Using the method described in [25], LMWHA-Fe_3O_4 NPs were prepared by dissolving 23 mg oleic acid-coated Fe_3O_4 NPs in 15 mL of toluene. LMWHA aqueous solution was additionally prepared by adding 25 mg LMWHA to 30 mL NaOH (1 mol/L) solution. The two solutions were placed in a reaction bottle and vigorously stirred for 24 h to replace the hydrophobic oleic acid on the Fe_3O_4 NP surface with hydrophilic LMWHA. After continuous rapid stirring for 24 h, the sample was left to stand for 20 min to allow the solution to separate. The light-colored liquid at the bottom of the solution was subjected to ultra-filtration and was then collected by centrifugation at 8000 rpm for 10 min. The pH of the solution was adjusted to 7 through the addition of 0.1 mol/L HCl solution and the excess water was then removed using a freeze dryer.

2.4. Determination of Ferrous Ion Content in Fabricated LMWHA-Fe_3O_4 NPs

The iron ion content in the fabricated LMWHA-Fe_3O_4 NPs was determined by reacting the iron ions with thiocyanate ions in a moderately acidic medium to form a dark red iron thiocyanate complex. In particular, 30% hydrochloric acid was added to the LMWHA-Fe_3O_4 NP samples at 55 °C for 3 h followed by the addition of ammonium persulfate for 15 min to form Fe^{3+} ions. Potassium thiocyanate

(KSCN) was then added to the solution. The SCN$^-$ ions reacted with the Fe^{3+} ions to form a blood-red colored complex in accordance with the following formula:

$$Fe^{3+}\ 6SCN^- \rightarrow [Fe(SCN)_6]^{3-}$$

The color intensity of the complex was determined by a microplate reader (EZ Read 400, Biochrom, Holliston, MA, USA) at a wavelength of 570 nm. The concentration of the iron oxide NPs in the prepared samples was then determined by comparing the measured color intensity with the intensity readings obtained for a series of standard solutions with known Fe^{3+} concentrations.

2.5. Characterization of Fabricated LMWHA-Fe$_3$O$_4$ NPs

The superparamagnetic properties of the fabricated LMWHA-Fe$_3$O$_4$ NPs were determined using a superconducting quantum interference device (SQUID) (MPMS7, Quantum Design, San Diego, CA, USA). The hysteresis loops of the LMWHA-Fe$_3$O$_4$ NPs were measured at temperatures of 5 K and 300 K. The saturation magnetizations of the oleic acid-coated Fe$_3$O$_4$ NPs and LMWHA-Fe$_3$O$_4$ NPs were also tested and compared.

The dynamic viscosities of the LMWHA and LMWHA-Fe$_3$O$_4$ NPs were measured at 25 °C using a viscometer (X-420, Cannon Instrument Co., State College, PA, USA). As described in a previous study [26], the samples were added to pure water to form solutions with a concentration of 0.5 mg/mL. The solutions were then stirred magnetically for 2 h and the dynamic viscosity was read with units of centistokes (cSt) using DD water as a control.

The thermal stabilities of the oleic acid-coated Fe$_3$O$_4$ NPs, neat LMWHA, and LMWHA-Fe$_3$O$_4$ NPs were measured using a thermogravimeter (TGA, TG 209 F3 Tarsus, Netzsch, Gerätebau GmbH, Bavarian, Germany). An amount of 5 mg of each sample was heated from room temperature to 700 °C at a rate of 10 °C/min in a chamber filled with nitrogen. The decomposition temperatures (T_d) and residual weights of the various samples at 700 °C were then measured and compared.

2.6. In Vitro Biocompatibility Tests of LMWHA-Fe$_3$O$_4$ NPs

MCF7 breast cancer cells were seeded onto Petri dishes with a density of 1×10^4 cells/mL and maintained in alpha modified Eagle's minimum essential medium (αMEM) supplemented with 4 mM L-glutamine, 10% fetal bovine serum, and 1% penicillin-streptomycin. The cells were incubated at 37 °C in a 5% CO$_2$ environment for periods of 3, 6, 9 and 12 h, respectively. The cytotoxicity of the prepared materials was evaluated in accordance with the ISO10993-5 standard [26]. Briefly, according to ISO 10993-5, LMWHA and LMWHA-Fe$_3$O$_4$ NPs were immersed in the cultured medium with a concentration of 0.2 g/mL at 37 °C for 24 h. Liquid extracts were collected and added to the culture medium of MCF7 cells with concentrations of 0.1, 0.2 and 0.4 mg/mL, respectively. MCF7 cells cultured with 2% dimethyl sulfoxide (DMSO) and cultured medium alone were used as positive and negative controls, respectively. After co-culturing the cells with the liquid extracts for 24 h, the cell viability was determined using the MTT (3-(4,5-Dimethylthiazol-2-yl)-2,5-diphenyltetrazolium bromide) method. In addition, the absorbance was determined at 570/690 nm wavelengths using a microplate reader (EZ Read 400, Biochrom, Holliston, MA, USA).

2.7. Determination of Binding Quantity of LMWHA-Fe$_3$O$_4$ NPs to MCF7 Cells

MCF7 cells were cultured in 6-well culture dishes with densities of 1×10^4, 5×10^4, 1×10^5 and 5×10^5 cells/mL, respectively. After the cells were attached to the plate, the medium was removed and replaced with a new medium containing 1 mg/mL of LMWHA-Fe$_3$O$_4$ NPs. After culturing the cells for an additional 12 h, the medium was aspirated and the cells were washed twice with PBS to remove the untargeting LMWHA-Fe$_3$O$_4$ NPs. An amount of 200 mL 0.05% (v/v) Triton X-100 was then added to each dish. After cell disruption by three freeze/thaw cycles, cell lysate from each dish was transferred to 1.5-mL microcentrifuge tubes. The thiocyanate method (see Section 2.4) was then used to measure the binding amount of LMWHA-Fe_3O_4 NPs to the MCF7 cells.

2.8. Time-of-Flight Secondary Ion Mass Spectrometry Analysis

MCF7 cells were cultured in 6-well culture dishes with a density of 5×10^5 cells/mL. After the cells were attached to the culture plate, the medium was removed and replaced by new media containing 1 mg/mL oleic acid-coated Fe_3O_4 NPs, LMWHA and LMWHA-Fe_3O_4 NPs, respectively. After co-culturing for 24 h, the media were aspirated and the cells were washed twice with PBS to remove the untargeting materials. The treated cells were fixed by glutaraldehyde treatment. Time-of-flight secondary ion mass spectrometry (TOF-SIMS) (PHI TRIFT IV, ULVAC-PHI, Kanagawa, Japan) was then used to evaluate the targeting status of the LMWHA-Fe_3O_4 NPs on the MCF7 cell surface [27]. In performing the TOF-SIMS process, the Bi_3^+ primary ion beam (operated at 30 keV) was supplied by a Bi liquid metal ion gun fitted to the instrument. The distributions of the iron ions and phosphocholine fragments in the phospolipid were identified by the *m/z* 56 and *m/z* 86 signals, respectively. In addition, secondary ion images were obtained by scanning the ion beam across the cell surface over an area of 200×200 μm.

2.9. Statistical Analysis

For the cell viability and iron concentration tests, the mean values and standard deviations of each measurement were recorded. One-way analysis of variance (ANOVA) with Tukey's post hoc (SPSS Inc., Chicago, IL, USA) tests were then performed to evaluate the differences between the samples. A *p*-value lower than 0.05 was considered to be statistically significant in every case.

3. Results and Discussion

3.1. Characterization Results for Fe_3O_4 NPs

In preparing Fe_3O_4 NPs, preventing particle aggregation and obtaining a good dispersibility is an important concern. This problem is commonly addressed by coating the surface of the NPs with some form of polymer [7]. The polymer helps the NPs bind to other substances, and hence inhibits their aggregation. However, many polymer coatings may exhibit cytotoxicity, and therefore suppress cell differentiation and may cause cell death and apoptosis [8]. Accordingly, in the present study, the Fe_3O_4 NPs were coated with biocompatible oleic acid. The acid not only prevents oxidation reaction, but also suppresses the aggregation of the NPs, and hence reduces their size [28]. Figure 1a presents a TEM image of the oleic acid-coated NPs. It can be seen that the NPs have a spherical shape and are well dispersed in the cultured medium. The results are thus consistent with those presented in a previous report for oleic acid-coated magnetite NPs [29]. As shown in Figure 1b, the NP particle size is distributed mainly (71%) in the range of 4 to 8 nm and is less than 20 nm in every case.

Figure 2 presents the X-ray diffraction pattern of the oleic acid-coated Fe_3O_4 NPs. The sharp diffraction peaks at $2\theta = 30.1°, 35.4°, 43.1°, 53.2°, 56.9°$ and $62.52°$, respectively, indicate that the NPs have an inverse spinel structure [30,31]. In other words, the success of the Fe_3O_4 NP synthesis process is confirmed.

3.2. Characterization Results for LMWHA-Fe_3O_4 NPs

The LMWHA prepared by 20 kGy γ-irradiation exposure was found to have a molecular weight of 230 kDa. In general, Fe_3O_4 NPs coated with oleic acid can only be dissolved in organic solvents (i.e., not in water). However, HA is extremely hydrophilic and is insoluble in organic solvents. Accordingly, a mixing problem occurs at the interface between the oleic acid-coated F_3O_4 NPs and the LMWHA. Previous studies have proposed two methods for overcoming this problem [25]. In the first method, poly (ethylene glycol), poly (ethylene oxide), or poly (vinyl alcohol) is used to modify the surface of the Fe_3O_4 NPs. In the second method, the chemical structure of the HA is modified in some way, e.g., the carboxyl group of HA can be used to reduce its hydrophilicity [32]. However, excessive modification may reduce the targeting ability of the HA to the CD44 receptors of the cancer cells. In 2014, Chan et al. reported that the oleic acid coated on Fe_3O_4 NPs can be easily replaced by a polymer having more carboxylic acid or phosphate functional groups [32]. Accordingly, in the present study, alkaline solution was used to accelerate the saponification reaction of the oleic acid. Once the saponification reaction is completed and the supernatant was removed, the LMWHA-Fe_3O_4 NPs formed.

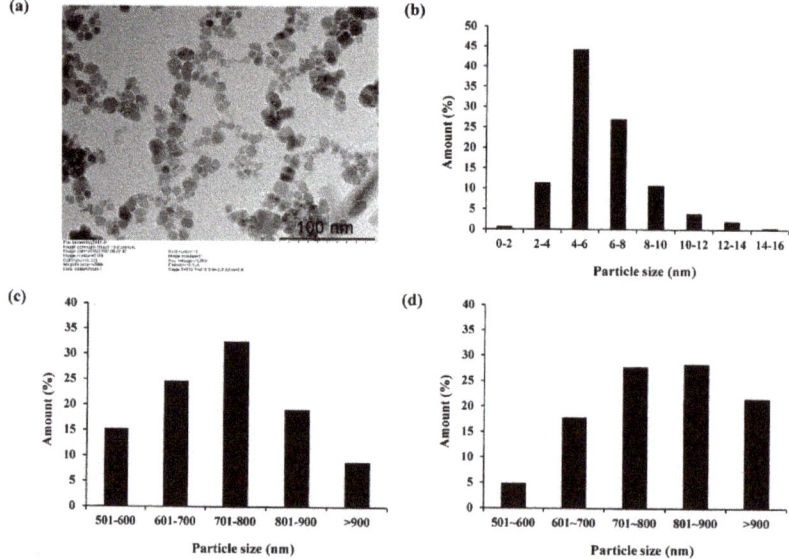

Figure 1. (a) Transmission electron microscopy (TEM) image of oleic acid-coated Fe_3O_4 NPs and (b) particle diameter distribution of nanoparticles; (c,d) particle diameter distributions of LMWHA and LMWHA-Fe_3O_4 NPs, respectively.

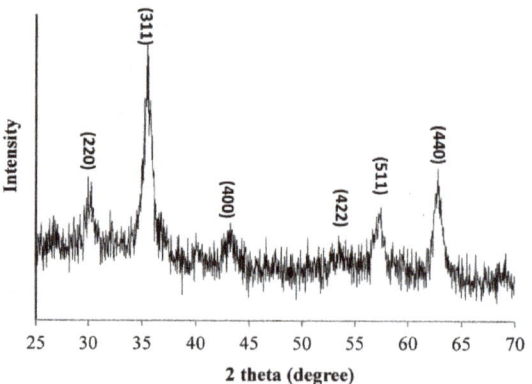

Figure 2. XRD pattern of oleic acid-coated Fe_3O_4 NPs.

Figure 1c,d shows the particle size distributions of the neat LMWHA gel and LMWHA-Fe_3O_4 NPs, respectively. As shown in Figure 1c, the LMWHA particle size falls mainly in the range of 600–900 nm. Referring to Figure 1d, the number of LMWHA-Fe_3O_4 NPs with a size larger than 700 nm (77.4%) is much higher than that of LMWHA particles with a similar size (60.2%). This finding is reasonable since the measured particle size in Figure 1d reflects the total value of HA and many iron oxide NPs [33].

The iron ion content of the LMWHA-Fe$_3$O$_4$ NPs was evaluated using the thiocyanate method (see Section 2.4). After quantifying the red ferric thiocyanate complex, the iron ion content was determined to be 4.43%.

The stability of the colloidal dispersions of the LMWHA-Fe$_3$O$_4$ NPs in water was evaluated by means of zeta potential measurements. The zeta potential of oleic acid-coated Fe$_3$O$_4$ NPs and LMWHA-Fe$_3$O$_4$ NPs were found to be −45.30 and −43.84 mV, respectively. These values were consistent with the value reported in the literature [34]. In general, the Zeta measurement result indicates that the LMWHA-Fe$_3$O$_4$ NPs have excellent colloidal stability due to a charge repulsion effect, which inhibits their aggregation in aqueous solutions.

3.3. Hysteresis Loop Detection

Previous studies have shown that Fe$_3$O$_4$ NPs with a diameter lower than 30 nm can pass through a superparamagnetic-ferromagnetic transition and exhibit superparamagnetic behavior [35,36]. As described in Section 3.1, the present Fe$_3$O$_4$ NPs have a diameter of less than 20 nm (see Figure 1b). Thus, it is reasonable to assume that they may exhibit a superparamagnetic behavior. However, previous studies have neither confirmed nor disproved the existence of such a phenomenon when the nano-particles are used to modify LMWHA. To address this gap, the present study investigated the magnetic properties of the prepared Fe$_3$O$_4$ NPs and LMWHA-Fe$_3$O$_4$ NPs using a quantum interference technique. As shown in Figure 3a, the saturation magnetization of the Fe$_3$O$_4$ NPs was found to exceed 60 emu/g when exposed to a high-strength magnetic field at 300 K. However, the saturation magnetization of the LMWHA-Fe$_3$O$_4$ NPs was only 5.9 emu/g due to the low concentration (4.43%) of Fe$_3$O$_4$ NPs in the complex. This finding is consistent with the results of previous studies, which also showed a reduction in the saturation magnetization of Fe$_3$O$_4$ NPs following modification with nature polymer [37,38]. However, despite the low Fe$_3$O$_4$ NP concentration ratio, the LMWHA-Fe$_3$O$_4$ NPs still exhibit a typical superparamagnetic property [36]. That is, a hysteresis loop is not observed at 300 K, but can be seen at 5 K (Figure 3b). Notably, this finding suggests that the LMWHA-Fe$_3$O$_4$ NPs have significant potential as an MRI contrast agent that can respond to a strong magnetic field [31].

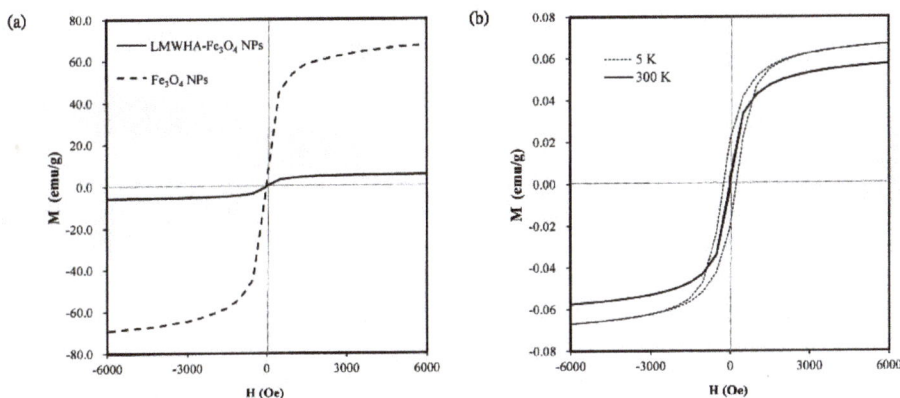

Figure 3. (a) Saturation magnetization of acid-coated Fe$_3$O$_4$ NPs and LMWHA-Fe$_3$O$_4$ NPs at temperature of 300 K. (b) Hysteresis loops of acid-coated Fe$_3$O$_4$ NPs and LMWHA-Fe$_3$O$_4$ NPs at 5 K and 300 K.

3.4. Viscosity Analysis

In practice, the high viscosity of HA limits its application as a blood vessel-injectable material into the human body. Furthermore, even though the viscosity of HA can be reduced by lowering its molecular weight, the efficiency of such an approach depends significantly on the particular method used. For example, while γ-irradiation and enzyme treatment can both reduce the molecular weight of HA, the depolymerization process and breaking site of the HA structure are different in each case. Huang et al. (2019) found that γ-irradiated HA exhibits a Newtonian liquid viscosity behavior due to the collapse of the macromolecular coils during the depolymerization process [17]. Figure 4 shows the measured kinematic viscosities of the present LMWHA and LMWHA-Fe_3O_4 NPs. It can be seen that even though an γ-irradiation process was used to fabricate the LMWHA, the dynamic viscosity of the LMWHA is still significantly higher than that of water. However, when the LMWHA is used to modify the Fe_3O_4 NPs, the kinematic viscosity decreases to just 1.27–3.00 cSt over the considered concentration range of 0.1–0.8 mg/mL. These values are very close to those of water. In 2012, Fakhari et al. tested the rheological behavior of HA/NP mixtures. Their results also showed that the addition of HA NPs can reduced the viscosity of HA to a value very close to those of water [39]. In other words, modification of the LMWHA by the Fe_3O_4 NPs results in a significant reduction in its viscosity and hence improves its potential as a blood vessel-injectable material. Although how NPs can control rheological properties of HA solutions is less well known, the mechanism whereby increasing the relative proportion of NPs in the solution reduces the viscosity of the HA/NP mixtures should be achieved by interrupting HA self-association. This is because the participating sites were occupied, and thus accessibility was reduced [39]. In this regard, it is speculated that the reduction in viscosity showed in Figure 4 is the result of an altered electrostatic attraction between the HA polymer and the Fe_3O_4 NPs, which reduces the intermolecular force between the HA and HA molecules.

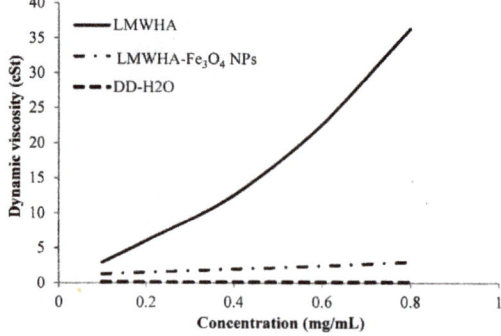

Figure 4. Dynamic viscosities of LMWHA-Fe_3O_4 NPs and neat LMWHA with various concentrations.

3.5. Thermogravimetric Analysis

Previous studies have reported that the addition of Fe_3O_4 NPs to polylactide acid improves the thermal stability of the polymer [40]. The TGA analysis results obtained in the present study show that the prepared LMWHA, Fe_3O_4 NPs and LMWHA-Fe_3O_4 NPs have decomposition temperatures (T_d) of 71.75 °C, 197.8 °C and 80.1 °C, respectively (see Figure 5a,b). In other words, the addition of the nano-Fe_3O_4 particles increases the thermal stability of the LMWHA polymer due to their high T_d temperature. In addition, the presence of the Fe_3O_4 NPs also increases the residual mass left after the composites have undergone thermogravimetric testing at 700 °C (see Figure 5c). In particular, the char residuals at the end of the TGA runs of the LMWHA, Fe_3O_4 NPs and LMWHA-Fe_3O_4 NPs are 25.82%, 79.41% and 63.94%, respectively. Notably, the thermal residual weight of the LMWHA-Fe_3O_4 NPs obtained in the present study is almost twice that reported in a previous study [25]. It is hence inferred that the Fe_3O_4 NPs prepared through the synthesis route proposed in the present study have a higher iron ion content, and are thus more suitable for MRI imaging and/or hyperthermia therapy.

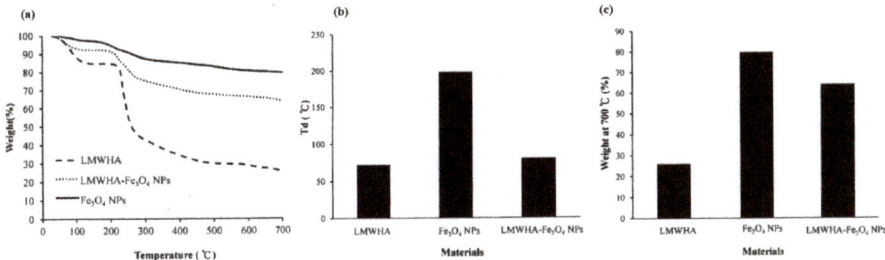

Figure 5. (a) Thermogravimetric (TGA) patterns of Fe_3O_4 NPs, LMWHA-Fe_3O_4 NPs and neat LMWHA. (b) Td values and (c) residual weights of Fe_3O_4 NPs, LMWHA-Fe_3O_4 NPs and neat LMWHA.

3.6. Cytotoxicity Characterization Results

Figure 6 shows the in vitro test results for the cytotoxicity of the prepared LMWHA and LMWHA-Fe_3O_4 NPs toward the MCF7 cells. As shown in Figure 6a, the viability of the MCF7 cells cultured with 2% DMSO is reduced by around 20% compared to the blank condition. However, there is no change in cell viability is found under co-culturing with liquid extracts of neat LMWHA and LMWHA-Fe_3O_4 NPs. No significant difference is observed in the viabilities of the MCF7 cells co-cultured with the neat LMWHA and LMWHA-Fe_3O_4 NPs, respectively, for the considered concentrations of less than 0.4 mg/mL. It was reported that the cytotoxicity of the Fe_3O_4 nanoparticles is greatly dependent on the particle size. Xie et al. (2016) investigated the cytotoxic effects Fe_3O_4 NPs with different diameters on the human hepatoma cells. Their results indicated that 6 nm Fe_3O_4 NPs exhibited negligible cytotoxicity. However, Fe_3O_4 NPs with particle size larger than 9 nm may affect cytotoxicity by inducing cellular mitochondrial dysfunction or impairing the integrity of plasma membrane [41]. The particle size of the prepared Fe_3O_4 NPs is mainly concentrated between 4 and 6 nm, and this is the reason the Fe_3O_4 NPs prepared in this study showed no cytotoxic effects on cells. Furthermore, for both materials, the cell viability increases over the considered 12 h culture period, as shown in Figure 6b. Figure 7 shows that no significant morphological change of the MCF7 cells occurs following co-culturing with liquid extracts of neat LMWHA and LMWHA-Fe_3O_4 NPs, respectively. Overall, the results presented in Figures 6 and 7 show that neither neat LMWHA nor the LMWHA-Fe_3O_4 NPs exert a cytotoxic effect on the MCF7 cells. Previous studies indicated that γ-irradiation-treated HA significantly showed improvement effect on the viability of fibroblasts [17,42]. However, a comparable improvement in the MCF7 cell viability was not observed (Figure 6b), as shown in previous studies [17,42]. The apparent discrepancy between the two sets of results most likely arises due to the cells were co-cultured with liquid extract rather than with the prepared material directly.

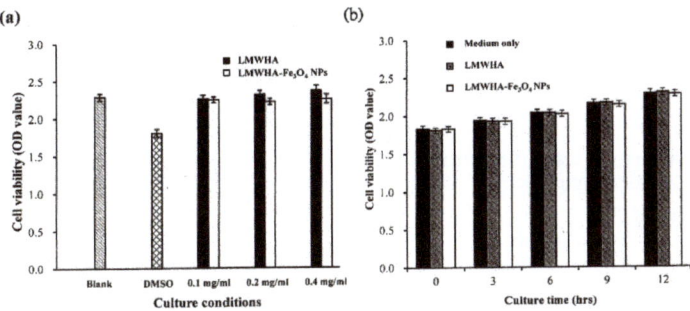

Figure 6. (a) Cytotoxicity test results, and (b) cell proliferation assay results for MCF7 cells cultured with neat LMWHA and LMWHA-Fe_3O_4 NPs.

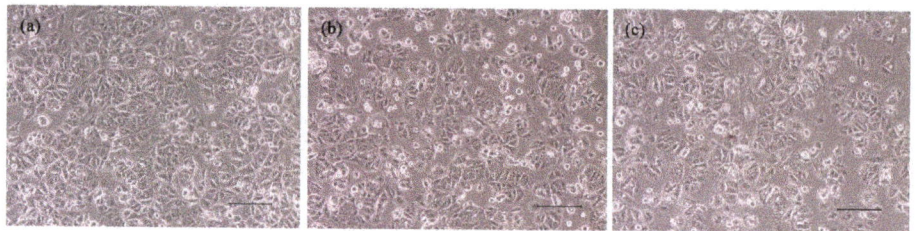

Figure 7. Morphologies of MCF7 cells cultured with (**a**) material-free pure medium, and media with (**b**) neat LMWHA and (**c**) LMWHA-Fe$_3$O$_4$ NPs. Scale bar denoted 100 μm.

3.7. Time-of-Flight Secondary Ion Mass Spectrometry Analysis

As shown in Figure 4, the dynamic viscosity property alternation of the prepared LMWHA-Fe$_3$O$_4$ NPs may also affect its targeting ability on the cell surface. However, this phenomenon has not been investigated or discussed previously. The Fe$_3$O$_4$ NPs targeting on the MCF7 cell surface were extracted to form a solution containing iron ions. The ion concentration was then determined using the thiocyanate colorimetry technique described in Section 2.4. As shown in Figure 8, the extracted quantity of Fe^{3+} ions increased significantly with increasing cell concentration. However, the iron ion amount does not seem to increase proportionally with the increase of cells number. This phenomenon may be due to the HA solved in water was not in a homogeneous status.

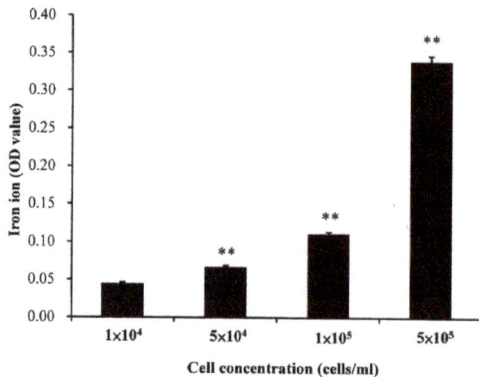

Figure 8. Quantity of iron ions on surface of MCF7 cells with different concentrations. ** $p < 0.01$.

It has been reported that TOF-SIMS is a highly sensitive and chemically specific analytical tool for both inorganic and organic subjects [43–45]. In the TOF-SIMS process, a high-energy ion beam (referred to as the primary ion beam) is applied to the sample and used to generate ionized molecular fragments from the sample surface. The ionized fragments (referred to as secondary ions) are then collected by a time-of-flight mass analyzer and separated according to their mass-to-charge ratio (m/z). The collected data are then used to reconstruct the chemical distribution of the sample surface [46]. For biological applications, TOF-SIMS is an effective technique for analyzing the chemical composition of cellular membranes with a few molecular layers in depth [44] and can identify the chemical change of phospholipid molecules on the surface of single cells or tissues [27,45–49]. Furthermore, TOF-SIMS can be used to analyze whether inorganic material or metal is incorporated in the extracellular matrix and to detect the uptake (or otherwise) of metal iron by human cells [43,45].

Figure 9a,b shows the TOF-SIMS signal intensity and image, respectively, of the Fe$_3$O$_4$ NPs (m/z 56) prepared in the present study. It is well known that phospholipids are the most abundant molecules on mammalian cellular membranes. The signal at m/z 86 represents a fragment (C$_5$H$_{12}$N$^+$)

of phosphocholine; a head group of the phospholipids [43,44]. Thus, in TOF-SIMS analysis, the m/z 86 image reflects the chemical composition of membranes [44]. As shown in Figure 10, a high-intensity m/z 86 signal exists in the central areas of the present MCF cells, as also described in a previous study [44]. Furthermore, no 56 signal is observed for the MCF7 cells cultured with medium, LMWHA or Fe_3O_4 NPs alone. However, for the MCF7 cells cultured with LMWHA-Fe_3O_4 NPs, a visible iron ion signal is found. Figure 11 shows the relations between the total ions, m/z 56 and m/z 86. The convolution image of m/z 56 and 86 indicates that the Fe_3O_4 NPs exist on the cellular surface.

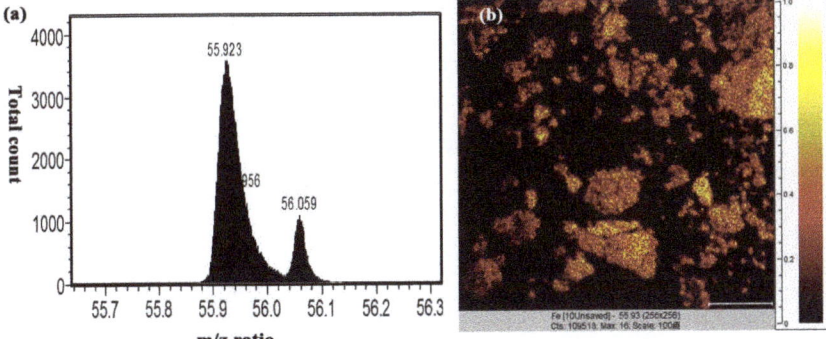

Figure 9. (a) m/z 56 signal and (b) TOF-SIMS image of fabricated Fe_3O_4 NPs.

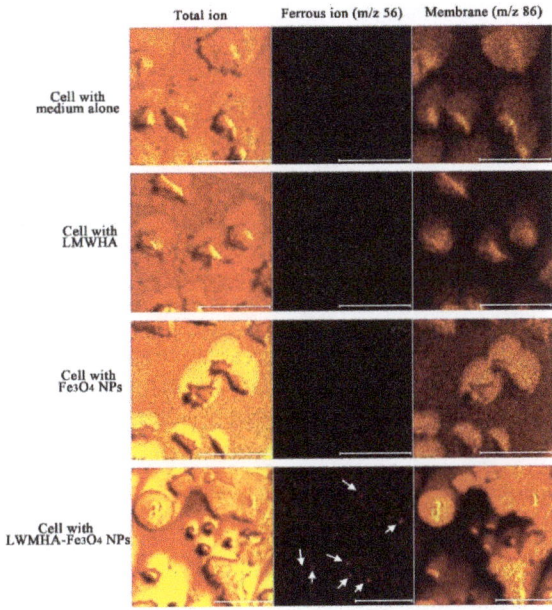

Figure 10. TOF-SIMS images of MCF7 cells cultured with medium alone, LMWHA, Fe_3O_4 NPs, and LMWHA-Fe_3O_4 NPs. Please note that the m/z 56 signal (red dots identified by white arrows) can be found only for cells cultured with LMWHA-Fe_3O_4 NPs. Note also that m/z 86 images represent cellular membrane.

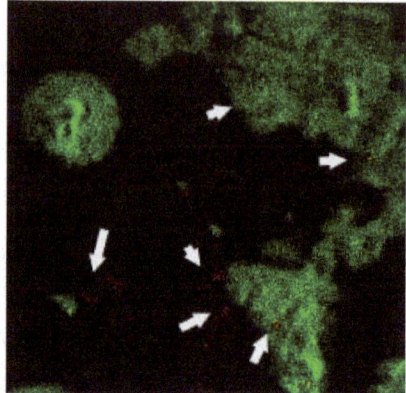

Figure 11. Convolution TOF-SIMS image consisting of *m/z* 56 signal (red image representing iron ion) and *m/z* 86 mass signal (green image representing membrane). Please note that Fe_3O_4 NPs (red dots identified by white arrows) are found to target MCF7 cellular surface.

Except for TOF-SIMS, scanning (SEM) and transmission (TEM) electron microscopy are techniques that can also provide information regarding the interactions between cells and the prepared LMWHA-Fe_3O_4 NPs. However, the sensitivity of the SEM is such that it is hard to observe a single ion. The limitation of this study is that only several layers of atoms can be observed using the TOF-SIMS technique. Thus, whether the prepared Fe_3O_4 NPs enter the cells cannot be observed as TEM imaging. However, the detection of biochemical responses due to the NPs enter the cell was not the purpose of this study. Thus, the conclusion of this study would not be changed even without TEM images. Overall, the results present in this study suggest that the prepared LMWHA-Fe_3O_4 NPs has significant potential to be developed as an injectable agent for targeting breast cancer tumors in biomedical application.

Author Contributions: Conceptualization, H.-M.H., H.-T.W.; Data curation, P.-C.C., P.-H.W. and C.-M.L.; Funding acquisition, K.-H.F., W.-J.C.; Investigation, S.-Y.L., Methodology, H.-T.W.; Project administration, H.-M.H. All authors have read and agreed to the published version of the manuscript.

Funding: This research received no external funding.

Conflicts of Interest: The authors declare no conflict of interest.

References

1. Major, J.L.; Meade, T.J. Bioresponsive, Cell-Penetrating, and Multimeric MR Contrast Agents. *Acc. Chem. Res.* **2009**, *42*, 893–903. [CrossRef]
2. Telgmann, L.; Sperling, M.; Karst, U. Determination of gadolinium-based MRI contrast agents in biological and environmental samples: A review. *Anal. Chim. Acta* **2013**, *764*, 1–16. [CrossRef]
3. Yang, C.-Y.; Tai, M.-F.; Chen, S.-T.; Wang, Y.-T.; Chen, Y.-F.; Hsiao, J.-K.; Wang, J.-L.; Liu, H.-M. Labeling of human mesenchymal stem cell: Comparison between paramagnetic and superparamagnetic agents. *J. Appl. Phys.* **2009**, *105*, 7. [CrossRef]
4. Gupta, A.K.; Gupta, M. Synthesis and surface engineering of iron oxide nanoparticles for biomedical applications. *Biomaterials* **2005**, *26*, 3995–4021. [CrossRef]
5. Shen, Y.F.; Tang, J.; Nie, Z.; Wang, Y.; Ren, Y.; Zuo, L. Preparation and application of magnetic Fe_3O_4 nanoparticles for wastewater purification. *Sep. Purif. Technol.* **2009**, *68*, 312–319. [CrossRef]
6. Turcheniuk, K.; Tarasevych, A.; Kukhar, V.P.; Boukherroub, R.; Szunerits, S. Recent advances in surface chemistry strategies for the fabrication of functional iron oxide based magnetic nanoparticles. *Nanoscale* **2013**, *5*, 10729. [CrossRef]

7. Cai, H.; An, X.; Cui, J.; Li, J.; Wen, S.; Li, K.; Shen, M.; Zheng, L.; Zhang, G.; Shi, X. Facile Hydrothermal Synthesis and Surface Functionalization of Polyethyleneimine-Coated Iron Oxide Nanoparticles for Biomedical Applications. *ACS Appl. Mater. Interfaces* **2013**, *5*, 1722–1731. [CrossRef]
8. Li, J.; Zheng, L.; Cai, H.; Sun, W.; Shen, M.; Zhang, G.; Shi, X. Facile One-Pot Synthesis of Fe$_3$O$_4$@Au Composite Nanoparticles for Dual-Mode MR/CT Imaging Applications. *ACS Appl. Mater. Interfaces* **2013**, *5*, 10357–10366. [CrossRef]
9. Fraser, J.R.E.; Laurent, T.C. Hyaluronan: Its nature, distribution, functions and turnover. *J. Intern. Med.* **1997**, *242*, 27–33. [CrossRef]
10. Toole, B.P. Hyaluronan-CD44 Interactions in Cancer: Paradoxes and Possibilities. *Clin. Cancer Res.* **2009**, *15*, 7462–7468. [CrossRef]
11. Kahmann, J.D.; O'Brien, R.; Werner, J.; Heinegård, D.; Ladbury, J.; Campbell, I.D.; Day, A.J. Localization and characterization of the hyaluronan-binding site on the Link module from human TSG-6. *Structure* **2000**, *8*, 763–774. [CrossRef]
12. Holmes, M.W.A.; Bayliss, M.T.; Muir, H. Hyaluronic acid in human articular cartilage. Age-related changes in content and size. *Biochem. J.* **1988**, *250*, 435–441. [CrossRef] [PubMed]
13. Mattheolabakis, G.; Milane, L.; Singh, A.; Amiji, M.M. Hyaluronic acid targeting of CD44 for cancer therapy: From receptor biology to nanomedicine. *J. Drug Target.* **2015**, *23*, 605–618. [CrossRef] [PubMed]
14. Kablik, J.; Monheit, G.; Yu, L.; Chang, G.; Gershkovich, J. Comparative Physical Properties of Hyaluronic Acid Dermal Fillers. *Dermatol. Surg.* **2009**, *35*, 302–312. [CrossRef] [PubMed]
15. Cowman, M.K.; Hittner, D.M.; Feder-Davis, J. 13C-NMR Studies of Hyaluronan: Conformational Sensitivity to Varied Environments†. *Macromolecules* **1996**, *29*, 2894–2902. [CrossRef]
16. Ke, C.; Sun, L.; Qiao, D.; Wang, D.; Zeng, X. Antioxidant acitivity of low molecular weight hyaluronic acid. *Food Chem. Toxicol.* **2011**, *49*, 2670–2675. [CrossRef]
17. Huang, Y.C.; Huang, K.Y.; Lew, W.Z.; Fan, K.H.; Chang, W.J.; Huang, H.M. Gamma-Irradiation-Prepared Low Molecular Weight Hyaluronic Acid Promotes Skin Wound Healing. *Polymers* **2019**, *11*, 1214. [CrossRef]
18. Zhang, H.; Li, J.; Sun, W.; Hu, Y.; Zhang, G.-F.; Shen, M.; Shi, X. Hyaluronic Acid-Modified Magnetic Iron Oxide Nanoparticles for MR Imaging of Surgically Induced Endometriosis Model in Rats. *PLoS ONE* **2014**, *9*, e94718. [CrossRef]
19. Li, J.; He, Y.; Sun, W.; Luo, Y.; Cai, H.; Pan, Y.; Shen, M.; Xia, J.; Shi, X. Hyaluronic acid-modified hydrothermally synthesized iron oxide nanoparticles for targeted tumor MR imaging. *Biomaterials* **2014**, *35*, 3666–3677. [CrossRef]
20. Zhong, L.; Liu, Y.; Xu, L.; Li, Q.; Zhao, D.; Li, Z.; Zhang, H.; Zhang, H.; Kan, Q.; Li, J.; et al. Exploring the relationship of hyaluronic acid molecular weight and active targeting efficiency for designing hyaluronic acid-modified nanoparticles. *Asian J. Pharm. Sci.* **2018**, *14*, 521–530. [CrossRef]
21. Chen, H.; Qin, J.; Hu, Y. Efficient Degradation of High-Molecular-Weight Hyaluronic Acid by a Combination of Ultrasound, Hydrogen Peroxide, and Copper Ion. *Molecules* **2019**, *24*, 617. [CrossRef] [PubMed]
22. Hokputsa, S.; Jumel, K.; Alexander, C.; Harding, S.E. A comparison of molecular mass determination of hyaluronic acid using SEC/MALLS and sedimentation equilibrium. *Eur. Biophys. J.* **2003**, *32*, 450–456. [CrossRef]
23. Choi, J.-I.; Kim, J.-K.; Kim, J.-H.; Kweon, D.-K.; Lee, J.-W. Degradation of hyaluronic acid powder by electron beam irradiation, gamma ray irradiation, microwave irradiation and thermal treatment: A comparative study. *Carbohydr. Polym.* **2010**, *79*, 1080–1085. [CrossRef]
24. Wang, H.-T.; Chan, Y.-H.; Feng, S.-W.; Lo, Y.-J.; Teng, N.-C.; Huang, H.-M. Development and biocompatibility tests of electrospun poly-l-lactide nanofibrous membranes incorporating oleic acid-coated Fe$_3$O$_4$. *J. Polym. Eng.* **2014**, *34*, 241–245. [CrossRef]
25. El-Dakdouki, M.H.; El-Boubbou, K.; Zhu, D.C.; Huang, X. A simple method for the synthesis of hyaluronic acid coated magnetic nanoparticles for highly efficient cell labelling and in vivo imaging†. *RSC Adv.* **2011**, *1*, 1449–1452. [CrossRef] [PubMed]
26. Huang, Y.-C.; Huang, K.-Y.; Yang, B.-Y.; Ko, C.-H.; Huang, H.-M. Fabrication of Novel Hydrogel with Berberine-Enriched Carboxymethylcellulose and Hyaluronic Acid as an Anti-Inflammatory Barrier Membrane. *BioMed Res. Int.* **2016**, *2016*, 1–9. [CrossRef]

27. Wu, M.-P.; Chang, N.-C.; Chung, C.-L.; Chiu, W.-C.; Hsu, C.-C.; Chen, H.-M.; Sheu, J.-R.; Jayakumar, T.; Chou, D.-S.; Fong, T.-H. Analysis of Titin in Red and White Muscles: Crucial Role on Muscle Contractions Using a Fish Model. *BioMed Res. Int.* **2018**, *2018*, 1–11. [CrossRef]
28. Kim, D.; Zhang, Y.; Voit, W.; Rao, K.; Muhammed, M. Synthesis and characterization of surfactant-coated superparamagnetic monodispersed iron oxide nanoparticles. *J. Magn. Magn. Mater.* **2001**, *225*, 30–36. [CrossRef]
29. Zhang, L.; He, R.; Gu, H. Oleic acid coating on the monodisperse magnetite nanoparticles. *Appl. Surf. Sci.* **2006**, *253*, 2611–2617. [CrossRef]
30. Sun, J.; Zhou, S.; Hou, P.; Yang, Y.; Weng, J.; Li, X.; Li, M. Synthesis and characterization of biocompatible Fe_3O_4 nanoparticles. *J. Biomed. Mater. Res. Part A* **2006**, *80*, 333–341. [CrossRef]
31. Zhao, H.; Saatchi, K.; Häfeli, U.O. Preparation of biodegradable magnetic microspheres with poly(lactic acid)-coated magnetite. *J. Magn. Magn. Mater.* **2009**, *321*, 1356–1363. [CrossRef]
32. Chan, N.; Laprise-Pelletier, M.; Chevallier, P.; Bianchi, A.; Fortin, M.-A.; Oh, J.K. Multidentate Block-Copolymer-Stabilized Ultrasmall Superparamagnetic Iron Oxide Nanoparticles with Enhanced Colloidal Stability for Magnetic Resonance Imaging. *Biomacromolecules* **2014**, *15*, 2146–2156. [CrossRef] [PubMed]
33. Cai, H.; Li, K.; Shen, M.; Wen, S.; Luo, Y.; Peng, C.; Zhang, G.; Shi, X. Facile assembly of Fe_3O_4@Au nanocomposite particles for dual mode magnetic resonance and computed tomography imaging applications. *J. Mater. Chem.* **2012**, *22*, 15110. [CrossRef]
34. Shukla, S.; Jadaun, A.; Arora, V.; Sinha, R.K.; Biyani, N.; Jain, V.K. In vitro toxicity assessment of chitosan oligosaccharidecoated iron oxide nanoparticles. *Toxicol. Rep.* **2015**, *2*, 27–39. [CrossRef]
35. Cheng, F. Characterization of aqueous dispersions of Fe_3O_4 nanoparticles and their biomedical applications. *Biomaterials* **2005**, *26*, 729–738. [CrossRef]
36. Ge, J.; Hu, Y.; Biasini, M.; Beyermann, W.P.; Yin, Y. Superparamagnetic magnetite colloidal nanocrystal clusters. *Angew. Chem. Int. Ed. Engl.* **2007**, *46*, 4342–4345. [CrossRef]
37. Sanjai, C.; Kothan, S.; Gonil, P.; Saesoo, S.; Sajomsang, W. Chitosan-triphosphate nanoparticles for encapsulation of super-paramagnetic iron oxide as an MRI contrast agent. *Carbohydr. Polym.* **2014**, *104*, 231–237. [CrossRef]
38. Tong, S.; Hou, S.; Zheng, Z.; Zhou, J.; Bao, G. Coating Optimization of Superparamagnetic Iron Oxide Nanoparticles for High T2Relaxivity. *Nano Lett.* **2010**, *10*, 4607–4613. [CrossRef]
39. Fakhari, A.; Phan, Q.; Thakkar, S.V.; Middaugh, C.R.; Berkland, C.J. Hyaluronic Acid Nanoparticles Titrate the Viscoelastic Properties of Viscosupplements. *Langmuir* **2013**, *29*, 5123–5131. [CrossRef]
40. Wang, H.-T.; Chiang, P.-C.; Tzeng, J.-J.; Wu, T.-L.; Pan, Y.-H.; Chang, W.-J.; Huang, H.-M. In Vitro Biocompatibility, Radiopacity, and Physical Property Tests of Nano-Fe_3O_4 Incorporated Poly-l-lactide Bone Screws. *Polymers* **2017**, *9*, 191. [CrossRef]
41. Liu, P.F.; Liu, D.; Cai, C.; Chen, X.; Zhou, Y.; Wu, L.; Sun, Y.; Dai, H.; Kong, X.; Xie, Y. Size-dependent cytotoxicity of Fe_3O_4 nanoparticles induced by biphasic regulation of oxidative stress in different human hepatoma cells. *Int. J. Nanomed.* **2016**, *11*, 3557–3570. [CrossRef] [PubMed]
42. Ergun, G.; Egilmez, F.; Yilmaz, S. Effect of reduced exposure times on the cytotoxicity of resin luting cements cured by high-power led. *J. Appl. Oral Sci.* **2011**, *19*, 286–292. [CrossRef] [PubMed]
43. Kokesch-Himmelreich, J.; Schumacher, M.; Rohnke, M.; Gelinsky, M.; Janek, J. ToF-SIMS analysis of osteoblast-like cells and their mineralized extracellular matrix on strontium enriched bone cements. *Biointerphases* **2013**, *8*, 17. [CrossRef] [PubMed]
44. Nygren, H.; Hagenhoff, B.; Malmberg, P.; Nilsson, M.; Richter, K. Bioimaging TOF-SIMS: High resolution 3D imaging of single cells. *Microsc. Res. Tech.* **2007**, *70*, 969–974. [CrossRef] [PubMed]
45. Pour, M.D.; Ren, L.; Jennische, E.; Lange, S.; Ewing, A.G.; Malmberg, P. Mass spectrometry imaging as a novel approach to measure hippocampal zinc. *J. Anal. At. Spectrom.* **2019**, *34*, 1581–1587. [CrossRef]
46. Jungnickel, H.; Laux, P.; Luch, A. Time-of-Flight Secondary Ion Mass Spectrometry (ToF-SIMS): A New Tool for the Analysis of Toxicological Effects on Single Cell Level. *Toxics* **2016**, *4*, 5. [CrossRef]
47. Brunelle, A.; Touboul, D.; Laprévote, O. Biological tissue imaging with time-of-flight secondary ion mass spectrometry and cluster ion sources. *J. Mass Spectrom.* **2005**, *40*, 985–999. [CrossRef]

48. Fletcher, J.S. Cellular imaging with secondary ion mass spectrometry. *Analyst* **2009**, *134*, 2204. [CrossRef]
49. Sjovall, P.; Johansson, B.; Lausmaa, J. Localization of lipids in freeze-dried mouse brain sections by imaging TOF-SIMS. *Appl. Surf. Sci.* **2006**, *252*, 6966–6974. [CrossRef]

© 2020 by the authors. Licensee MDPI, Basel, Switzerland. This article is an open access article distributed under the terms and conditions of the Creative Commons Attribution (CC BY) license (http://creativecommons.org/licenses/by/4.0/).

Article

The Mechanical Properties, Secondary Structure, and Osteogenic Activity of Photopolymerized Fibroin

Ivan Bessonov [1,2,†], Anastasia Moysenovich [1,†], Anastasia Arkhipova [1,3], Mariam Ezernitskaya [4], Yuri Efremov [5], Vitaliy Solodilov [6], Peter Timashev [5,6], Konstantin Shaytan [1], Alexander Shtil [7,8] and Mikhail Moisenovich [1,*]

1. Biological Faculty, Lomonosov Moscow State University, 119234 Moscow, Russia; ivanbessonov@gmail.com (I.B.); a-moisenovich@mail.ru (A.M.); anastasia-yu.arkhipova@yandex.ru (A.A.); shaytan49@yandex.ru (K.S.)
2. JSC Efferon, 143026 Moscow, Russia
3. Regional Research and Clinical Institute ("MONIKI"), 129110 Moscow, Russia
4. A. N. Nesmeyanov Institute of Organoelement Compounds, Russian Academy of Sciences, 119334 Moscow, Russia; ezernits@mail.ru
5. Institute for Regenerative Medicine, Sechenov University, 119991 Moscow, Russia; yu.efremov@gmail.com (Y.E.); timashev.peter@gmail.com (P.T.)
6. Semenov Institute of Chemical Physics Russian Academy of Sciences, 119991 Moscow, Russia; vital-yo@yandex.ru
7. Blokhin National Medical Research Center of Oncology, 115478 Moscow, Russia; shtilaa@yahoo.com
8. Institute of Gene Biology, Russian Academy of Sciences, 119991 Moscow, Russia
* Correspondence: mmoisenovich@mail.ru
† These authors contributed equally to this work.

Received: 1 February 2020; Accepted: 10 March 2020; Published: 12 March 2020

Abstract: Previously, we have described the preparation of a novel fibroin methacrylamide (FbMA), a polymer network with improved functionality, capable of photocrosslinking into Fb hydrogels with elevated stiffness. However, it was unclear how this new functionality affects the structure of the material and its beta-sheet-associated crystallinity. Here, we show that the proposed method of Fb methacrylation does not disturb the protein's ability to self-aggregate into the stable beta-sheet-based crystalline domains. Fourier transform infrared spectroscopy (FTIR) shows that, although the precursor ethanol-untreated Fb films exhibited a slightly higher degree of beta-sheet content than the FbMA films (46.9% for Fb-F-aq and 41.5% for FbMA-F-aq), both materials could equally achieve the highest possible beta-sheet content after ethanol treatment (49.8% for Fb-F-et and 49.0% for FbMA-F-et). The elasticity modulus for the FbMA-F-et films was twofold higher than that of the Fb-F-et as measured by the uniaxial tension (130 ± 1 MPa vs. 64 ± 6 MPa), and 1.4 times higher (51 ± 11 MPa vs. 36 ± 4 MPa) as measured by atomic force microscopy. The culturing of human MG63 osteoblast-like cells on Fb-F-et, FbMA-F-et-w/oUV, and FbMA-F-et substrates revealed that the photocrosslinking-induced increment of stiffness increases the area covered by the cells, rearrangement of actin cytoskeleton, and vinculin distribution in focal contacts, altogether enhancing the osteoinductive activity of the substrate.

Keywords: silk fibroin; methacrylated silk fibroin; tissue engineering; photocrosslinking; osteogenic differentiation

1. Introduction

The mechanical characteristics of scaffolds for tissue engineering are of the utmost importance for cell attachment and differentiation. Cells on the scaffold apply mechanical forces, thereby causing its deformation. In turn, elastic deformation of the substrate provides cytoskeletal tension. Focal contacts initiate the signaling from the surface to the cytoplasm and the nucleus. On the rigid substrate resistant to tension, the cells tend to spread, whereas they are largely round-shaped on soft and easily deformable supports [1,2]. Thus, the substrate's elasticity modulus (EM) can regulate cell morphology, motility, and the direction of differentiation. The *EM* of normal bone extracellular matrix (ECM) is 20–50 kPa [3], whereas for the mineralized bone, the values are as large as 10–20 GPa [4]. The respective values for typical polymer scaffolds are in the kPa range [5]; an increase in *EM* can be a factor of the osteogenic potency of the material [6–8].

Fibroin (Fb) is a perspective source of scaffolds for tissue engineering due to its remarkable biocompatibility, moderate immunomodulating activity, bioresorbability, yield of non-toxic products (mostly amino acids and short peptides), and its mechanical strength [7]. In particular, the fibroin-based materials are well-described scaffolds with perspective osteogenic properties [9–11]. In a dry state, the *EM* values of Fb fibers are within the GPa range [12]. However, for porous swollen scaffolds, these parameters normally decrease to 5–25 kPa [13]. This fact can be explained by the specific primary and secondary structure of fibroin [14]. Namely, the repetitive sequences of the hydrophobic amino acid residues assemble into beta sheets that, in turn, form nanoscale crystalline domains within the polymer. Therefore, the Fb-based materials are semicrystalline. The content, localization, and orientation of crystalline domains may depend on the method of material fabrication; therefore, the mechanical properties would be different [15]. Fb modifications at the side groups are promising for synthesis of polymers by an additional functional network [16,17] with optimized characteristics.

Chemically modified Fb substrates have been demonstrated to be promising materials for bone regeneration [17]. In this study, a urethane methacrylate Fb derivative was prepared; photopolymerization of this macromonomer yielded covalently cross-linked films. The degree of cross-linking negatively correlated with *EM* values and beta-sheet crystallinity. The urethane methacrylic Fb derivative [17] has been synthesized using a highly reactive isocyanoethyl methacrylate that provided an exhaustive modification of abundant serine residues (~12% of total amino acid content) in the hydrophobic repeats (GAGAGS) that form the crystalline domains of beta sheets. We have proposed another synthetic route for Fb methacrylation, yielding the methacrylamide derivative FbMA [18]. In doing so, we introduced the methacrylic moieties under mild conditions selectively at the amino groups of Arg and Lys residues, whose total presence in Fb is only ~0.5%. Expectedly, new sparse covalent bonds in the photochemically cross-linked FbMA gel were formed in protein's amorphous domains. As the amount, size, orientation, and distribution of crystalline domains are essential for superior mechanical characteristics of the material, photocrosslinking via the amorphous domains is beneficial to maintain the remarkable Fb properties. Thus, FbMA, being a convenient parental compound for the photochemical fabrication (including SLA and DLP additive manufacturing) of covalently cross-linked Fb gels, can provide the scaffolds with a higher EM. This property is important for osteogenic potency and makes FbMA preferable over the materials described in [17].

In this study, we report that FbMA-based materials retain the pristine fibroin's beta-sheet crystallinity and its properties as a scaffold for cell culture. Importantly, the elevated stiffness of FbMA-based materials provides an osteogenic response of human osteoblast-like cells (MG63 line).

2. Materials and Methods

2.1. Reagents

Methacrylic anhydride (94%), Na_2CO_3, LiBr, and diphenyl-(2,4,6-trimethylbenzoyl)phosphine oxide (TPO, 97%) were purchased from Sigma-Aldrich (Darmstadt, Germany). 1,1,1,3,3,3-Hexafluoro-2-propanol (HFIP, 99%) was purchased from P&M–Invest (Moscow, Russia), and ethanol (95%) was purchased from Medchimprom (Moscow Region, Balashikha, Russia).

2.2. Isolation of Fb

Lyophilized Fb was obtained from surgical silk threads LLC «Optikum» (Moscow, Russia) using the established protocols [19]. Briefly, threads were boiled for 30 min in aqueous 0.02 M Na_2CO_3, and then rinsed for 3 × 30 min in distilled water to remove Na_2CO_3. The threads were dried in the oven at 60 °C for 4 h, and then dissolved in 9.3 M aqueous LiBr at 60 °C for 3–4 h. The viscous yellow solution was dialyzed for 2 days in deionized water. The resulting solution was frozen for 2 days, vacuum dried at −20 °C until complete sublimation, and then stored in ambient conditions in closed vials.

2.3. Synthesis of FbMA

Methacrylated fibroin (FbMA) was synthesized according to a procedure described by us [18]. Briefly, 1 g of lyophilized Fb was placed in a round-bottom flask equipped with a magnetic stirrer and 20 mL of 0.1 M potassium phosphate buffer solution (pH 7.0) was added. Fb was dissolved on a water bath at 50 °C under continuous stirring to reach the final protein concentration of 5% (w/v). One milliliter of methacrylic anhydride was then added to the mixture. The reaction was continued under stirring at 50 °C for 1 h, followed by the addition of 20 mL 0.1 M potassium phosphate buffer solution (pH 7.0) and cooling the reaction mixture to room temperature. The solution was dialyzed against a 20-fold volume of deionized H_2O through a cellulose cut-off dialysis tubing under constant stirring until the smell of methacrylic anhydride disappeared; water was exchanged with a fresh portion every hour. The resulting product was placed in a Petri dish and frozen at −18 °C. Freeze-drying up to constant mass afforded FbMA as a white powder (yield 97%).

2.4. Preparation of Fb-HFIP Solution

Forty milligrams of lyophilized Fb was added to a glass vial with a screw cap and equipped with a magnetic stirrer. Four milliliters of HFIP was added to the vial. Fibroin and HFIP were kept in a closed vial under a continuous stirring for 2 h at 50 °C, until a clear yellow viscous solution was obtained. The solution was stored at ambient temperature before the films were cast.

2.5. Preparation of FbMA-HFIP Solution

Forty milligrams of powdered FbMA were added to an amber glass vial with a screw cap and equipped with magnetic stirrer. Two milligrams of TPO photoinitiator and 4 mL of HFIP were added, and the vials were kept in the dark. FbMA, TPO, and HFIP were kept in the closed vial under continuous stirring for 2 h at 50 °C until a clear yellow viscous solution was obtained. The prepared solution was stored at ambient temperature before the films were cast.

2.6. Film Fabrication

Six-hundred microliters of 10 mg/mL Fb/HFIP solution was pipetted on a smooth surface of an open-type mold (flat injection-molded polypropylene sheet) and left for 2 h at room temperature under the fume hood. As the solvent evaporated, a transparent film was detached from the mold and placed into 50 mL of sterile 0.9% aqueous NaCl. The solution was changed every 8 h for 48 h. The washed film was stored in a tightly closed vessel under sterile 0.9% NaCl solution at 4 °C.

The Fb-F-aq film was placed into 70% aqueous ethanol for 2 h, and then into 96% ethanol for 24 h. The dehydrated films were kept under 96% ethanol in a tightly closed vessel at 4 °C.

To obtain FbMA-F-aq, 600 µL of 10 mg/mL FbMA/HFIP solution was pipetted on a smooth surface of an open-type mold (flat injection-molded polypropylene sheet) and left for 2 h at room temperature under the fume hood. The dry film was formed as described above. Then, the film was irradiated with 365 nm UV light at an intensity of 10 mW/cm^2 for 10 min and placed into 50 mL of sterile 0.9% aqueous NaCl. The solution was changed every 8 h for 48 h. The washed film was stored in a tightly closed vessel under sterile 0.9% NaCl solution at 4 °C.

The FbMA-F-aq film was placed into 70% aqueous ethanol for 2 h, and then into 96% ethanol for 24 h. The dehydrated films were kept under 96% ethanol in a tightly closed vessel at 4 °C.

FbMA-F-et-w/oUV were obtained as described above, but the UV irradiation step was omitted.

2.7. Fourier Transform Infrared Spectroscopy (FTIR)

Fourier transform infrared spectroscopy (FTIR) spectra were obtained on a Vertex 70 V Fourier spectrometer (Bruker, Ettlingen, Germany) using an ATR accessory with a diamond crystal (Pike, Fitchburg, WI, USA); the ATR spectra were collected in vacuum after 128,256 scans over a range of 4000 to 400 cm^{-1} with a resolution of 4 cm^{-1}. All corrections were done using an Omnic 8 program package.

2.8. Mechanical Tests

To determine the mechanical characteristics of studied materials, T-bone-shaped test specimens were obtained from the films by applying the cutting method. A series of 5 test specimens were cut for each type of material. Sample dimensions: thickness of 40 µm, width of 6 mm, and bearing length of 33 mm. One day prior to testing, Fb-F-et and FbMA-F-et samples were transferred from ethanol solution into an excessive volume of 0.9% NaCl.

The thin paper soaked in 0.9% NaCl was stuck to the surface to prevent drying during the test. As soaking disrupts the paper to individual fibers, its effect on mechanical characteristics of the films was negligible. Uniaxial tension was performed on a Zwick Z 100 test machine at a load rate 10 mm/min. Stress–strain curves (σ–ε) were registered during the loading. These curves were used to determine (according to ASTM D882-12) conditional yield strength, σ_T; at conditional elongation, ε_T; tensile strength, σ_p; elastic modulus, E, and elongation at break, ε_P. Standard deviations were calculated for each value.

2.9. Atomic Force Microscopy (AFM)

Nanomechanical measurements were performed on a Bioscope Resolve Atomic Force Microscopy (AFM) (Bruker, Santa Barbara, CA, USA) system mounted on an Axio Observer inverted optical microscope (Carl Zeiss, Jena, Germany). The ScanAsyst-Fluid cantilevers were used with the nominal spring constant 0.7 N/m and the nominal tip radius 20 nm. The exact values were determined by thermal tune method and by scanning of the titanium roughness sample (Bruker), respectively. All AFM measurements were performed in phosphate-buffered saline at room temperature. Force volume mapping was performed over 10 × 10 µm areas with 16 × 16 force curve arrays. Three samples of each film were analyzed (3–6 random areas per sample). The vertical piezo movement speed was 2 µm/s, and the trigger force was ≈25 nN. Each force curve was analyzed to extract the Young's modulus value using the NanoScope Analysis software (Bruker) with the Hertz model [20]:

$$F = \frac{4}{3}\frac{E}{1-v^2}\sqrt{R}\delta^{\frac{3}{2}}$$

where F is the measured force, E is the Young's modulus, v is the Poisson ratio (assumed to be 0.5), R is the tip radius, and δ is the indentation depth.

2.10. Culturing Human MG63 Osteoblast-Like Cells on Fb Substrates

The Fb-F-et, FbMA-F-et-w/oUV, and FbMA-F-et films were sterilized in 70% ethanol overnight, washed thrice with Eagle-MEM containing 1% nonessential amino acids (NEAA), then five times with Eagle-MEM supplemented with 10% fetal bovine serum (FBS) and 1% NEAA (30 min each washing) at 37 °C, 5% CO_2, transferred to Petri dishes with fresh medium and left overnight at 37 °C, 5% CO_2. The medium was discarded, 2 mL of MG63 (ATCC® CRL1427™) cell suspension (2.3 × 10^4 cells) in Eagle-MEM supplemented with 10% FBS and 1% NEAA. Then, cells were incubated overnight at 37 °C, 5% CO_2. By day 7, in culture the medium was changed for an osteogenic medium containing α-MEM with 5% FBS, 1% NEAA, 1% β-glycerophosphate, 0.01% dexamethasone, and 0.1% ascorbic acid, and left in the incubator for 14 days. The medium was changed every 3 days.

2.11. Cell Viability Tests

By days 1, 4, and 7 of culturing MG63 cells on Fb-F-et, FbMA-F-et-w/oUV, and FbMA-F-et surface cell viability was assessed by reducing 3-(4,5-dimethylthiazol-2-yl)-2,5-diphenyltetrazolium bromide (MTT tests). Films with cells were transferred into Petri dishes containing 2 mL of serum-free Eagle-MEM and 625 µg/mL MTT reagent, and incubated for 4 h at 37 °C, 5% CO_2. Then films were placed in DMSO to dissolve formazan, centrifuged at 16,800 g for 5 min. The supernatant was transferred into a 96-well plate. Optical density was measured at 550 nm. Films without cells were similarly processed as a control.

2.12. Cytoskeleton Morphology

By day 1 of culturing on Fb-F-et, FbMA-F-et-w/oUV and FbMA-F-et the MG63 cells were fixed with 4% paraformaldehyde in saline for 30 min at room temperature in the dark followed by three washings in saline. Then, the cells were permeabilized with 0.1% Triton X-100 in saline supplemented with 0.1% FBS for 10 min at 4 °C and washed twice with saline/0.1% FBS. Non-specific protein binding was blocked with PBS/1% FBS/0.1% Tween-20. To visualize focal contacts, cells were treated with mouse anti-vinculin antibody (ThermoFisher Scientific, Waltham, MA, USA, 1:200 in PBS/0/1% FBS/0/1% Tween-20; 1 h). After washing the secondary CF™ 543, conjugated anti-mouse IgG (H + L) (Sigma-Aldrich; Darmstadt, Germany; 1:700) was added. To visualize the actin cytoskeleton, cells were incubated with phalloidin conjugated with Alexa488 (Sigma-Aldrich, Darmstadt, Germany) as recommended by the manufacturer, and washed three times with saline. Hoechst 33342 ((ThermoFisher Scientific™, Waltham, MA, USA; 1 µg/mL) was added to counterstain the nuclei. Samples were analyzed on a Nikon Ti-E microscope with a confocal module A1 and the objective Apo TIRF Plan Fluor 63 × 1.49.

2.13. Activity of Alkaline Phosphatase (ALP)

The activity of Alkaline Phosphatase (ALP) was determined at days 7 and 14 of culture of MG63 cells on Fb-F-et, FbMA-F-et-w/oUV, and FbMA-F-et films in the osteogenic medium. Cells on the films were lysed in the buffer (50 mM Tris base, 100 mM glycin and 0.1% Triton X-100, XH 10.5). One milliliter of colorless p-nitrophenylphosphate (10 mg/mL), an ALP substrate, was added. Lysates were placed in the dark and incubated for 30 min. The reactions were stopped with 0.5 mL 0.2 M NaOH. ALP hydrolyzes p-nitrophenylphosphate to form a yellow p-nitrophenol. The activity of ALP was assessed by optical density at 405 nm. Values of ALP activity were normalized to the values in MTT tests. Films without cells were used as a control.

2.14. Deposition of Calcium Phosphate by MG63 Cells on Fb Scaffolds

By day 14 of culturing MG63 cells on Fb-F-et, FbMA-F-et-w/oUV, and FbMA-F-et under osteogenic conditions, samples were fixed in 2.5% glutaraldehyde, washed three times with deionized water, placed into 1.5 mL of 2% solution of alizarin red (pH 4.1–4.3), and incubated in the dark for 1 h at room temperature. Then, samples were washed 5 times with water. Calcium salts were examined on an inverted microscope Axiovert 200 M, Plan-Neofluar 20×/0.5 objective and AxioCam MRC 5 camera (Carl Zeiss, Jena, Germany). For quantitative analysis, samples stained with alizarin red were put into 1 mL of 10% CPC solution for 1 h. The optical density of the supernatant was measured at 540 nm and normalized by MTT values. Films without cells were used for control.

3. Results

3.1. Film Fabrication

The experimental scheme of fibroin films preparation is shown in Figure 1.

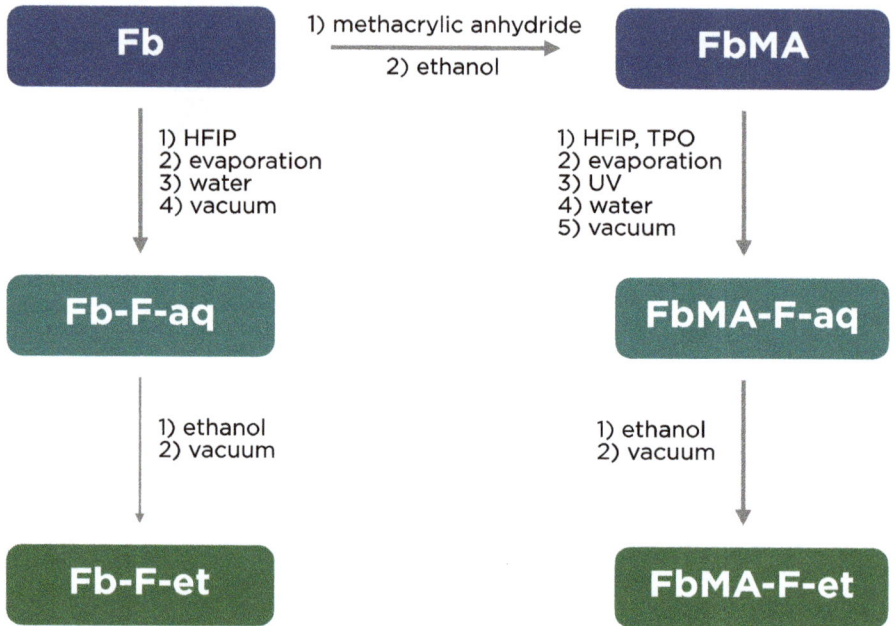

Figure 1. Preparation of fibroin (Fb) films.

Pristine Fb as a lyophilisate was used for synthesis of its methacrylated derivative FbMA [18]. To fabricate the films, these proteins were dissolved in HFIP. The Fb-F-aq film was fabricated from pristine Fb solution in HFIP by solvent evaporation and subsequent treatment with water. The FbMA-F-aq film was fabricated in a similar manner, but before its transfer into water, the film was subjected to UV irradiation in the presence of photoinitiator TPO to form covalent carbon–carbon bonds by the methacrylic residues. FbMA-F-et and Fb-F-et films were fabricated upon treatment of the corresponding water-treated precursor films with an excess of ethanol.

3.2. Mechanical Properties

Figure 2 shows that both Fb-F-et and FbMA-F-et were deformed elastically, i.e., the stress arisen in the material was directly proportional to elongation. Notably, the linear sections of the diagrams do not coincide. However, the curve of Fb-F-et is steeper. Further loading led to irreversible deformation. The bend of the loading diagram corresponds to the material's yield strength, which is the same for both materials.

Table 1 summarizes the mechanical characteristics of the films. The conditional yield strength σ_T is lower than the tensile strength σ_p, although these values are within the same range. This indicates that, during the tension process, the films become more stiff. The orientation of crystalline domains in an amorphous matrix further increases the strength of the films. EM values differed twofold: 64 MPa for Fb-F-et compared to 130 MPa for FbMA-F-et. The yield strength for FbMA-F-et was achievable at 4% extension vs. 9% for Fb-F-et.

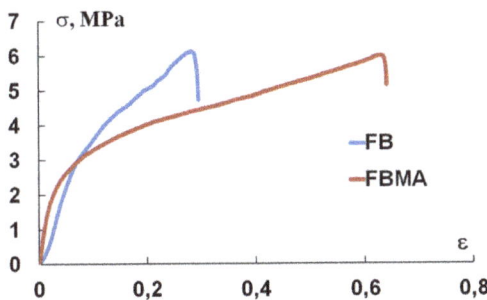

Figure 2. Stress–strain curve of Fb-F-et and FbMA-F-et films.

Table 1. Mechanical parameters of Fb-F-et and FbMA-F-et films.

Parameter Sample	Conditional Yield Strength σ_T, MPa	Tensile Strength σ_p, MPa	EM, MPa	Conditional Elongation, ε_T at σ_T	Elongation, ε_p at σ_p
FbMA-F-et	2.8 ± 0.5	5.8 ± 0.5	130 ± 1	0.4 ± 0.1	0.61 ± 0.12
Fb-F-et	3.5 ± 0.4	6.9 ± 0.8	64 ± 6	0.9 ± 0.2	0.43 ± 0.14

Also, the mechanical properties of the films were assessed by AFM nanoindentation experiments. The films were attached to the surface. Unlike uniaxial tension, which provides information about mechanics on a macro-scale (bulk properties), AFM analyses the properties specifically in the surface layer on a 10–100 nm scale. This scale is especially relevant considering the size of the structures responsible for the cell–surface interaction (focal adhesions), that is, hundreds of nm to several μm [21]. Examples of AFM nanomechanical maps are shown in Figure 3. Overall, the FbMA-F-et film showed a significantly higher (51 ± 11 MPa), but also more heterogeneous, Young's modulus than Fb-F-et film (36 ± 4 MPa).

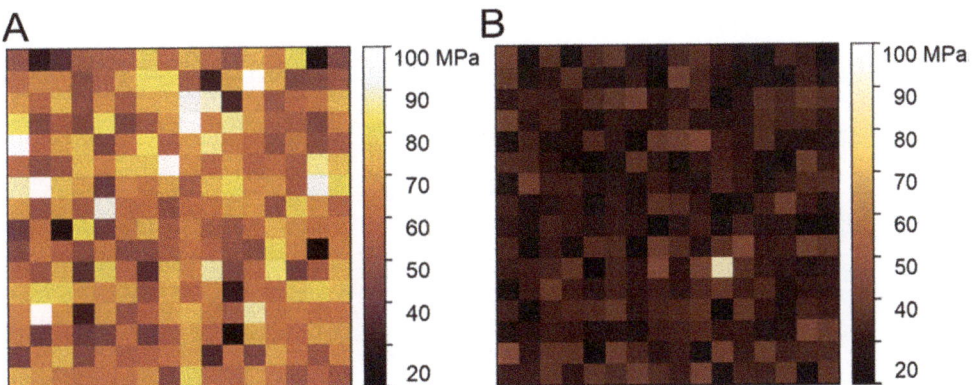

Figure 3. Nanomechanical maps (Young's modulus distributions) over a 10 × 10 μm area mapped using the Atomic Force Microscopy (AFM) Force Volume mode. (**A**) Representative map over the FbMA-F-et film. (**B**) Representative map over the Fb-F-et film. The color-coded scale of the Young's modulus is the same for both maps to better represent the difference. Note a significantly larger and more heterogeneous Young's modulus of FbMA-F-et film compared to Fb-F-et film.

3.3. Secondary Structure

FTIR spectroscopy was used to evaluate the degree of beta-sheet content (which correlates with crystallinity) of the materials. The shape of Amid I broad band (1590–1710 cm^{-1}) in the IR spectra of

Fb provides information about the secondary structure of the protein [22]. Figure 4 depicts the Amid I spectral range for all four films.

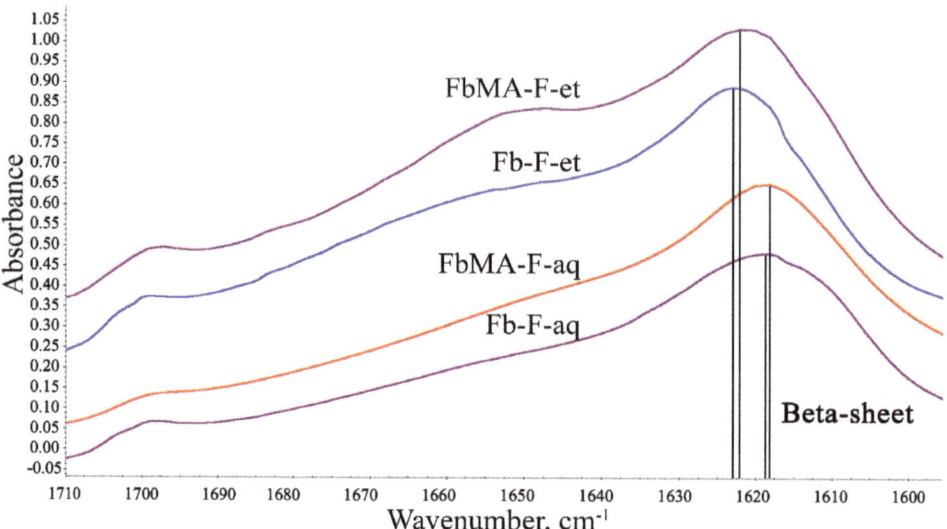

Figure 4. IR spectra of Fb-F-aq, FbMA-F-aq, Fb-F-et, and FbMA-F-et. Double lines show main peaks of beta sheets.

A broad Amid I band at ~1600–1700 cm^{-1} summarizes the individual stretching modes of amide groups of the secondary structures within the protein [23,24]. This band was resolved into individual components whose integral intensities were normalized to total intensity of the Amid I band. As the degree of beta-sheet content is a key determinant of Fb mechanical properties, the components in the ranges 1610 to 1635 cm^{-1} and 1695 to 1710 cm^{-1} assigned to β-sheets were the most important for interpretation. A general view of the respective spectral region with individual bands that correspond to the secondary structures of Fb-F-et is shown in Figure 5. Integral intensities of the peaks were used for quantitation of the impact of each structure (Table 2).

Table 2. FTIR spectra and secondary structures of aqueous Fb and FbMA films.

Secondary Structure	Wavenumber, cm^{-1}	FbMA-F-aq	Fb-F-aq	FbMA-F-et	Fb-F-et
Side chains	1590–1605	10.8	6.5	2.8	3.2
Beta-sheets	1610–1635, 1695–1710	41.5	46.9	49.8	49.0
Random coils	1635–1645	10.8	16.0	9.8	9.9
Beta-turns	1647–1654	12.3	15.1	18.2	18.1
Alpha-helixes	1658–1664	13.8	9.4	11.9	12.3
Bends and turns	1666–1695	10.8	6.2	7.4	7.5

Table 2 presents data for Fb-F-et and FbMA-F-et and their precursors Fb-F-aq and FbMA-F-aq, respectively. For Fb-F-aq, the total content of beta sheets was 46.9%, whereas for FbMA-F-aq, this value was slightly lower (41.5%). Other types of secondary structures also differed between the materials. The content of the side chains, alpha-helixes, bends, and turns was greater for FbMA-F-aq, whereas random coils and beta-turns were more frequent in Fb-F-aq films. However, Fb-F-et and FbMA-F-et treated with ethanol had similar percentages of beta-sheets: 49.0% and 49.8%, respectively. These two materials possess a maximal degree of beta-sheet content. No substantial differences between other types of secondary structures were observed after treatment with ethanol.

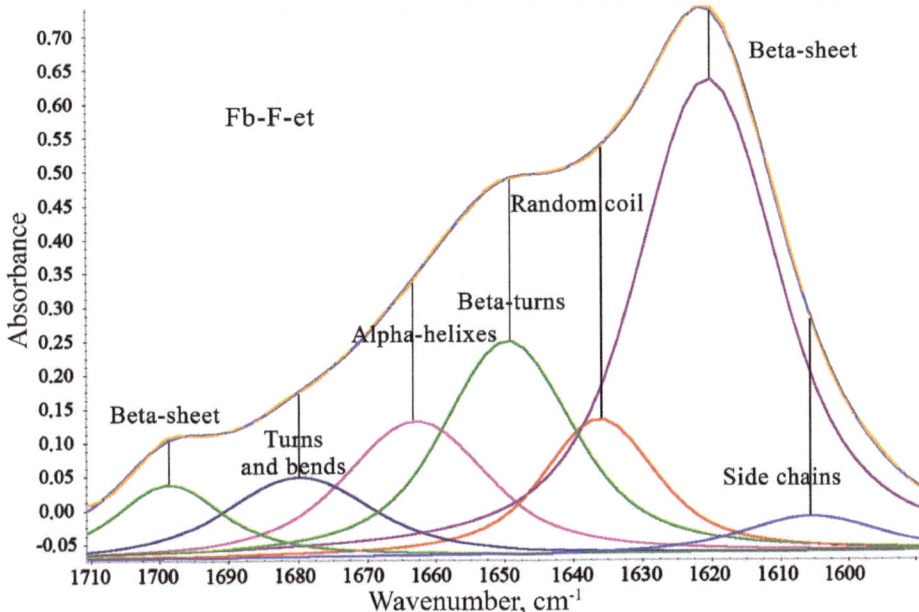

Figure 5. Individual bands that comprise the Amid I band in IR spectrum of Fb-F-et. Two bands at 1610–1635 cm^{-1} and 1695–1710 cm^{-1} correspond to the beta sheet structures.

3.4. Biological Properties

Films made of Fb-F-et and FbMA-F-et were used for culturing the MG63 cells. As shown in Figure 6A–C, no statistically significant changes in cell number, survival, and spreading were detectable between the cells grown on each substrate ($p > 0.05$).

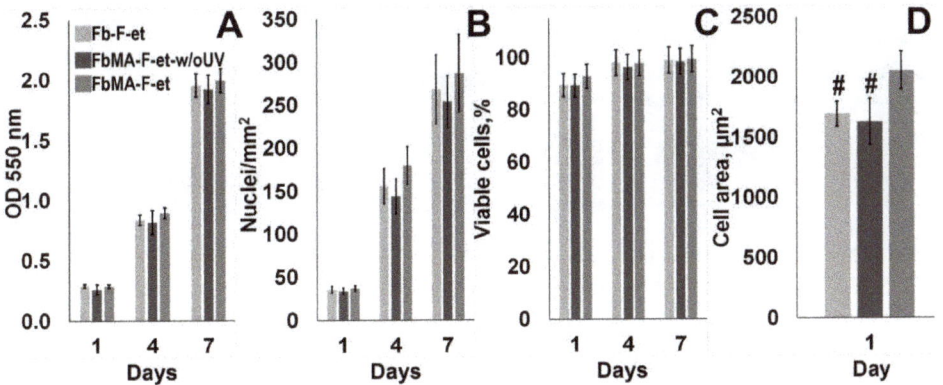

Figure 6. Cytocompatibility of FbMA-F-et. (**A**) Cell survival (MTT tests) at days 1, 4, and 7 of culture; (**B**) Number of nuclei per 1 mm^2; (**C**) % viable cells by LIVE/DEAD tests. (**D**) Cell area at one day of cultivation. #: statistically significant difference with FbMA-F-et group ($p < 0.05$).

A comparison of focal contacts and actin microfilaments in MG63 cells revealed that on Fb-F-et and FbMA-F-et-w/oUV F-actin was less ordered, and the number of stress fibers was smaller than on FbMA-F-et. FbMA-F-et films increased the cell surface compared to Fb-F-et and FbMA-F-et-w/oUV: 2003.75 ± 157.35 μm^2 vs. 1692.35 ± 103.74 μm^2 and 1627.15 ± 193.54, respectively (Figure 6D, $p < 0.05$). Furthermore, in cells grown on FbMA-F-et, vinculin was detectable in the cytoplasm, in focal contacts, as well as in the perinuclear area. In the cells cultured on Fb-F-et and FbMA-F-et-w/oUV, the amount of vinculin in the cytoplasm was rather small, and its structure in the focal contacts was less organized (Figure 7).

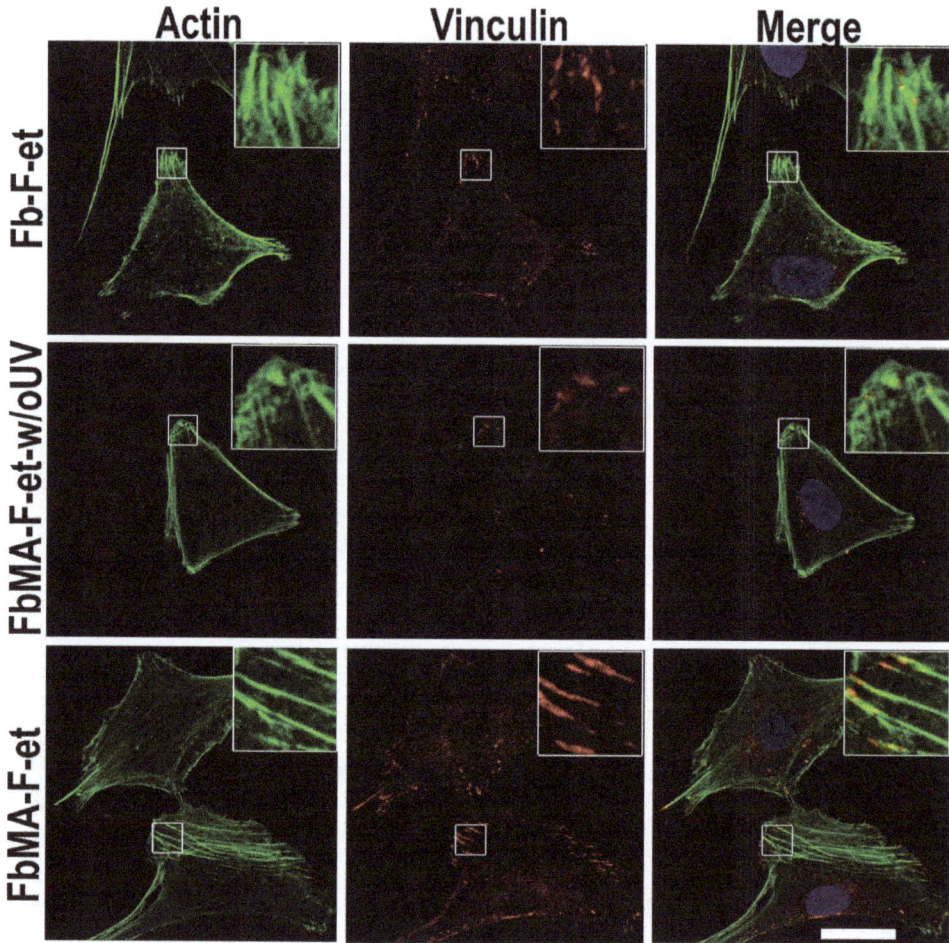

Figure 7. Stress fibers and vinculin organization of MG63 cells after 24 h on Fb-F-et, FbMA-F-et-w/oUV, or FbMA-F-et. Reconstructions are based on optical sections of cells growing on Fb-F-et (top), FbMA-F-et-w/oUV (middle), and FbMA-F-et (bottom). The cytoskeletal network was detected with Alexa488-phalloidin (green), and nuclei were stained with Hoechst 33,342 (blue). Vinculin (red) was visualized by immunocytochemistry. Scale bar, 25 μm.

Next, we determined ALP (an early marker of osteogenic activity of osteoblasts) and mineralization of Fb-F-et, FbMA-F-et and FbMA-F-et-w/oUV surfaces by MG63 cells. Maximum ALP activity was detectable by day 7; FbMA-F-et was more potent than Fb-F-et and FbMA-F-et-w/oUV (0.808 ± 0.027 vs.

0.721 ± 0.028 and 0.691 ± 0.047), respectively; $p < 0.05$). By day 14, ALP activity decreased in all groups (0.334 ± 0.031 vs. 0.388 ± 0.043 and 0.367 ± 0.037, respectively; $p > 0.05$; Figure 8A). At day 14, alizarin red staining in all groups was positive, indicating the deposition of calcium (Figure 8 right panel). On FbMA-F-et, calcium deposition was more pronounced compared to Fb-F-et and FbMA-F-et-w/oUV (OD values 1.772 ± 0.081 vs. 1.061 ± 0.074 and 0.994 ± 0.17, respectively; $p < 0{,}05$) (Figure 8B).

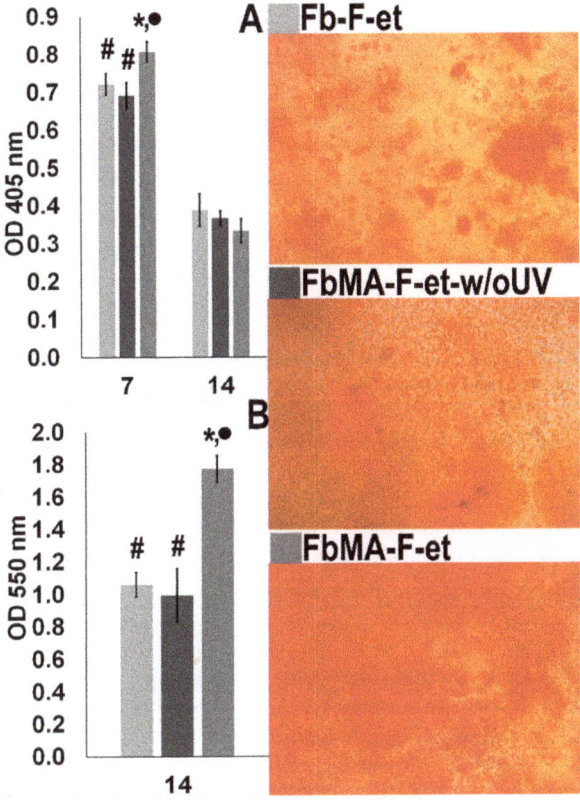

Figure 8. Osteogenic properties of Fb-derived substrates. (**A**) ALP activity by days 7 and 14 after induction of osteogenesis. (**B**) Quantitation of alizarin red staining after 14 days. Right panel: mineral depositions after 14 days of culturing MG63 cells under osteogenic conditions (staining with alizarin red). Statistically significant difference with * Fb-F-et group ($p < 0.05$); • FbMA-F-et-w/oUV ($p < 0.05$); # FbMA-F-et ($p < 0.05$).

4. Discussion

4.1. Film Fabrication

In this study, we compared the structure and characteristics of two types of 2D fibroin films (Figure 1). Fibron films are an important type of scaffolds for soft [25] and bone tissue regeneration [26]. The control samples are Fb-F-aq and Fb-F-et films based on the pristine Fb. These materials were fabricated by standard methods (dissolution in HFIP, solution casting, solvent evaporation, and optional ethanol treatment). HFIP, a highly efficient organic solvent for fibroin solubilization, has been used for fabrication of advanced protein-based materials [27]. FbMA-F-aq and FbMA-F-et films were fabricated via TPO-initiated photochemical polymerization of the novel methacrylated Fb derivative

FbMA. TPO is a monoacylphosphine-type, highly efficient photoinitiator that undergoes homolytic C–P bond cleavage under UV irradiation [28]. In situ generated free radicals initiate chain polymerization of methacrylic moieties of FbMA, that is, the solid cast FbMA film. Usually, Fb-based materials fabricated without a dehydrating step, such as alcohol treatment, are amorphous; their *EM* values are low («silk I»). Treatment with methanol or ethanol induced protein transition to a crystalline state and provided the material with a higher *EM* value («silk II»). This method of crystallization is critical for making rigid Fb scaffolds for bone regeneration. FbMA-F-aq specimens, unlike Fb-F-aq, possessed an additional network of covalent bonds that limit the mobility of polymer chains. The roles of methacrylic substituents and the cross-links in FbMA-based materials are important for the completion of crystallization and might even negatively influence this process [17]. This factor could alter their conformational transitions to beta-sheets, thereby preventing the formation of the crystalline structure in FbMA-F-et films. On the contrary, similar beta-sheet content for both ethanol-treated samples (Fb-F-et and FbMA-F-et) would indicate that the protein's self-assembly into a crystalline structure remained unaltered.

4.2. Mechanical Properties

The most dramatic difference between FbMA-F-et and Fb-F-et is *EM*: 130 ± 1 MPa vs. 64 ± 6 MPa, respectively (Table 1). Cross-linked FbMA-F-et film also showed a twofold lower elongation (0.4 vs. 0.9) (Table 1). We attributed this differential elasticity to the structure of new polymers (Figure 2). Additional covalent cross-links in FbMA-F-et limit the mobility of peptide chains in the swollen polymer. These findings differ from structurally similar materials reported elsewhere [17]. Using dry specimens, the authors have shown that *EM* decreased from 1.5 GPa for the non-cross-linked Fb to 0.5 GPa for the cross-linked counterpart. Thus, our method of Fb functionalization has no negative influence on mechanical characteristics. Therefore, photochemical cross-linking is relevant for scaffold preparation, so as this process does not decrease the rigidity of the material.

The higher *EM* values of FbMA-F-et film were confirmed by AFM: the film was ~40% stiffer than the Fb-F-et film (51 ± 11 MPa vs. 36 ± 4 MPa, respectively) (Figure 3). These values, however, were lower than those measured with the uniaxial tension. Such discrepancy is observed when macroscale and microscale measurements on the polymer samples are compared; one explanation is a difference between the properties of bulk and surface layers [29]. The higher local heterogeneity of FbMA-F-et films (Figure 3A) could be due to distribution of additional covalent cross-links [30]. Thus, covalent cross-linking of FbMA-F-et did not alter the strength and also increased the *EM* values. This elevated rigidity makes FbMA-F-et similar to the osseous tissue, further implicating this material for bone repair.

4.3. Secondary Structure

Conversion of silk I into silk II and quantification of crystallinity can be monitored by FTIR spectroscopy (Figures 4 and 5) [31,32]. HFIP-based Fb-F-aq and FbMA-F-aq films prepared without ethanol treatment exhibit lowered beta-sheet content on comparison to their ethanol-treated counterparts Fb-F-et and FbMA-F-et (Table 2). A beta-sheet content of 46.9% for Fb-F-aq is in an agreement with previously reported 48% for HFIP-based porous fibroin scaffolds [33]. These values are within the range of beta-sheet content achievable with alcohol treatment or similar dehydration step. These results indicate that the Fb-F-aq film is prone to spontaneous crystallization (prompted by vacuum at ATR accessory and the contact with residual NaCl from saline solution).

Chemically modified FbMA-F-aq, indeed, exhibits lowered beta-sheet content—41.5% (Table 2). In contrast to pristine fibroin, the covalently cross-linked methacrylic groups in FbMA-F-aq provide steric hindrance for conformational transition.

Treatment of these films with ethanol yields Fb-F-et and FbMA-F-et films. Despite the structural differences caused by covalent cross-links at the methacrylic substituents in the side chains, Fb-F-et and FbMA-F-et specimens showed similar degrees of beta-sheet content (49.0% vs. 49.8%; Table 2), as determined by FTIR spectra. Neither the introduction of methacrylic groups nor the novel –C–C–

bond formation has deteriorated beta-sheet formation. We attributed this result to the repetitive Ala/Gly sequences that are not targeted by methacrylation; these sequences are found in Fb domains critical for beta-sheet formation, so the Fb-to-FbMA transition is not associated with the changes of amino acid residues responsible for assembly into a secondary structure. In contrast, the hydrophilic domains enriched in Lys and Arg residues (into which the methacrylic groups were introduced) do not participate in the formation of the crystalline structure. One may suggest that the observed ~50% degree of beta-sheet content is maximal for «silk II»-type structure prepared by ethanol treatment. This value corresponds to the reported 51.2% beta-sheet content of Fb screws for orthopedic applications fabricated using intermediate HFIP solution and subsequent alcohol treatment [34]. Therefore, the observed differences in mechanical parameters such as EM, deformation, water contact angle, and swelling, are most likely due to the new covalent bonding rather than differential secondary structure. These results disagree with previous reports [17] in which the degree of beta-sheet content of photopolymerized Fb derivatives decreased along with the increased degree of cross-linking. We tend to explain this discrepancy by a specific composition of FbMA polymer. Indeed, in this material, the methacrylic residues were introduced into the amorphous domains, whereas in [17], the authors modified the serine residues involved in the formation of crystalline domains. As expected, direct methacrylation of serine residues prevented self-assembly of crystalline domains, a process dependent on GAGAGS repeats.

Consequently, crystallization of FbMA-F-aq in mild conditions (ambient temperature and low pressure at ATR accessory) is achievable with a lesser conversion than ethanol-induced crystallization. Harsh conditions, such as the prolonged treatment with excess of ethanol, provide the specimens with the same beta sheet content (Fb-F-et and FbMA-F-et). In other words, the covalently cross-linked FbMA-F-aq is less prone to spontaneous beta sheet formation compared to pristine Fb-F-aq. Nevertheless, both Fb-F-et and FbMA-F-et can achieve similar degrees of beta-sheet content at the thermodynamically equilibrium state.

4.4. Biological Properties

Materials made of native Fb are known to be highly biocompatible; moreover, these materials are applicable for bone regeneration [7,35,36] and fracture fixation [34,37]. These factors are associated with large mechanical strength and the ability to imitate the anionic charge of non-collagen proteins in the bone ECM, as well as with nucleation of hydroxyapatite and integrin mediated cell adhesion [38]. However, the rigidity of native silk is smaller compared to that in [4], so introduction of methacrylic moieties makes the material similar to the osseous tissue. Our study showed that the MG63 cells can attach, proliferate, and survive on FbMA-F-et films, indicating a good cytocompatibility of the scaffold (Figure 6A–C) [18,39]. Cell–substrate interactions are crucial for a plethora of physiological processes, including cell growth and differentiation, tissue remodeling, regeneration, viability, immune responses, etc. [40,41]. Physical parameters of the solid substrate important for the above interactions are topography, hydrophilicity, and rigidity [42]. As the geometry and the roughness of our materials were similar, differences in cell behavior can be attributable to the viscosity, rigidity, and hydrophilicity of the supports.

The increased cell area has been considered as a hallmark of osteogenic differentiation. Substrates that promote cell spreading are known to enhance osteogenesis, whereas the scaffolds on which the spreading is limited to retard this process [43]. The increased cell spreading is associated largely with dynamic polymerization of actin, that is, the formation of filamentous forms and subsequent restructuring into oriented stress fibers. Importantly, inhibition of actin polymerization attenuates the osteogenic differentiation [44]. We found that the area covered by the cells on the more rigid FbMA-F-et substrate was larger, with longer actin stress fibers along the cell bodies (Figures 6D and 7). In contrast, in the cells cultured on the softer Fb-F-et, actin was diffuse and amorphous although the stress fibers were visible across the cell body (Figure 7). Consequently, the markers of osteogenic differentiation such as ALP and calcium deposition were more pronounced on FbMA-F-et than on Fb-F-et. FbMA-F-et-w/oUV exerted the same effect as Fb-F-et.

Vinculin, a major component of focal contacts, is important for mechanosensing and signal transduction. Deregulation of vinculin is critical for cell adhesion, motility, and proliferation [45]. In particular, vinculin binds actin, thereby stimulating its polymerization and attracting the actin remodeling proteins [46]. Fb-F-et and FbMA-F-et evoked differential effects on morphology of MG63 cells. FbMA-F-et substrate provided formation of mature adhesion contacts at the periphery. Vinculin distribution on Fb-F-et was less organized than on FbMA-F-et (Figure 7). These results corroborate other findings suggesting that the cells adjust their shape and cytoskeleton to the stiffness of the substrate [47]. Interestingly, ALP activity and matrix mineralization were substantially more pronounced on FbMA-F-et (Figure 8). Therefore, in agreement with other studies [48], morphological changes are paralleled by biochemical markers of osteogenesis.

Introduction of methacrylic groups into the FbMA macromonomer occurs at the polar charged amino groups of Lys and Arg residues. The surface of FbMA-F-et is less polar, as the water contact angle is smaller (71° for FbMA-F-et vs. 60° for Fb-F-et) [18]. Furthermore, neutralization of the positive charge in the amino groups by methacrylation is expected to attenuate the net electric charge of the polymer. The roles of the charge and hydrophobicity of the substrate in ECM mineralization and ALP activity, as well as vinculin distribution in osteoblastic MC3T3-E1 cells on self-assembled monolayers of alkanethiols on gold with surface functionalization by –CH3, –OH, –COOH, and –NH_2 moieties, have been addressed in [49,50]. Cells cultured on hydrophilic surfaces showed elevated ALP activity and ECM mineralization. Also, genes specific for osteogenesis were upregulated compared to the cells cultured on the substrates functionalized by alkyl and carboxylic groups [36]. Both negative and positive charges promote the recruitment of vinculin during focal adhesion assembly in comparison with neutral and hydrophobic surfaces. Moreover, vinculin clusterization positively correlated with the presence of polar groups in the substrate [49]. One might expect a decreased vinculin in focal contacts and an attenuated ALP activity and mineralization on more hydrophobic substrates; however, we did not observe these phenomena. One explanation is that the effects of the increased rigidity prevailed over the effects of the lowered charge and surface hydrophilicity. On the other hand, hydrophobicity and the lowered charge can be compensated by the ability of Fb to adsorb positively charged hydrophilic molecules in the plasma, e.g., fibronectin and laminin, thereby altering the physico-chemical properties of the surface and countering the effects of hydrophobicity and negative charge on cellular functions [51].

5. Conclusions

Silk fibroin is one of the most prominent biomaterials for tissue regeneration; its complicated processability into the scaffolds of complex shape might be a major hindrance for its clinical applications. Previously, we have described a new mild method of Fb functionalization of its hydrophilic amorphous domains into the photocrosslinkable macromonomer FbMA. This approach to augment fibroin functionality accommodates it to SLA and DLP additive manufacturing technologies. This study provides further mechanical, structural, and biological evaluation of fabricated materials. Our findings indicate that the proposed method of Fb transformation yields the materials with a twice higher elasticity modulus, keeping the protein's beta-sheet crystallinity unaltered. FbMA-based materials retain the remarkable properties of Fb as a cell culture substrate, while enhancing their intrinsic osteoinductive properties via mechanical cues.

Author Contributions: Conceptualization, M.M.; validation, K.S., A.S., and P.T.; Supervision, M.M. and K.S.; investigation, M.M., A.M., A.A., M.E., Y.E., and V.S.; writing—original draft preparation, I.B., Y.E., A.M., and M.M.; writing—review and editing, P.T., K.S. and A.S.; visualization, A.M. and A.A. All authors have read and agreed to the published version of the manuscript.

Funding: This research was funded by the Russian Foundation for Basic Research, complex research project No. 17-00-00359, including research, No. 17-00-00357 (cells based experiments), No. 17-00-00358 (films preparation, cells based experiments), and by Russian academic excellence project "5–100" (experiments on AFM).

Conflicts of Interest: The authors declare no conflicts of interest. The funders had no role in the design of the study; in the collection, analyses, or interpretation of data; in the writing of the manuscript; or in the decision to publish the results.

References

1. Tee, S.Y.; Bausch, A.R.; Janmey, P.A. The mechanical cell. *Curr. Biol.* **2009**, *19*, 745–748. [CrossRef] [PubMed]
2. Sell, S.A.; Wolfe, P.S.; Garg, K.; McCool, J.M.; Rodriguez, I.A.; Bowlin, G.L. The use of natural polymers in tissue engineering: A focus on electrospun extracellular matrix analogues. *Polymers* **2010**, *2*, 522–553. [CrossRef]
3. Engler, A.J.; Sen, S.; Sweeney, H.L.; Discher, D.E. Matrix Elasticity Directs Stem Cell Lineage Specification. *Cell* **2006**, *126*, 677–689. [CrossRef] [PubMed]
4. Human, I.; Lamellae, B.; Dry, U. Nanoindentation Discriminates the Elastic Properties of. *Bone* **2002**, *30*, 178–184.
5. Yusupov, V.I.; Khmelenin, D.N.; Koroleva, A.; Volkov, V.V.; Asadchikov, V.E. Digging deeper: Structural background of PEGylated fibrin gels in cell migration and lumenogenesis. *RSC Adv.* **2020**, *10*, 4190–4200.
6. Bonartsev, A.P.; Bonartseva, G.A.; Reshetov, I.V.; Kirpichnikov, M.P.; Shaitan, K.V. Application of polyhydroxyalkanoates in medicine and the biological activity of natural poly(3-hydroxybutyrate). *Acta Nat.* **2019**, *11*, 4–16. [CrossRef]
7. Nguyen, T.P.; Nguyen, Q.V.; Nguyen, V.; Le, T.; Le, Q. Van Silk Fibroin-Based Biomaterials for Biomedical Applications: A Review. *Polymers* **2019**, *11*, 1933. [CrossRef]
8. Hixon, K.R.; Eberlin, C.T.; Kadakia, P.U.; McBride-Gagyi, S.H.; Jain, E.; Sell, S.A. A comparison of cryogel scaffolds to identify an appropriate structure for promoting bone regeneration. *Biomed. Phys. Eng. Express* **2016**, *2*, 035014. [CrossRef]
9. Bhattacharjee, P.; Kundu, B.; Naskar, D.; Kim, H.-W.; Maiti, T.K.; Bhattacharya, D.; Kundu, S.C. Silk scaffolds in bone tissue engineering: An overview. *Acta Biomater.* **2017**, *63*, 1–17. [CrossRef]
10. Correia, C.; Bhumiratana, S.; Yan, L.-P.; Oliveira, A.L.; Gimble, J.M.; Rockwood, D.; Kaplan, D.L.; Sousa, R.A.; Reis, R.L.; Vunjak-Novakovic, G. Development of silk-based scaffolds for tissue engineering of bone from human adipose-derived stem cells. *Acta Biomater.* **2012**, *8*, 2483–2492. [CrossRef]
11. Mandal, B.B.; Grinberg, A.; Seok Gil, E.; Panilaitis, B.; Kaplan, D.L. High-strength silk protein scaffolds for bone repair. *Proc. Natl. Acad. Sci. USA* **2012**, *109*, 7699–7704. [CrossRef] [PubMed]
12. Gosline, J.M.; Guerette, P.A.; Ortlepp, C.S.; Savage, K.N. The mechanical design of spider silks: From fibroin sequence to mechanical function. *J. Exp. Biol.* **1999**, *202*, 3295–3303. [PubMed]
13. Bai, S.; Han, H.; Huang, X.; Xu, W.; Kaplan, D.L.; Zhu, H.; Lu, Q. Silk scaffolds with tunable mechanical capability for cell differentiation. *Acta Biomater.* **2015**, *20*, 22–31. [CrossRef] [PubMed]
14. Zhou, C.Z.; Confalomieri, F.; Medina, N.; Zivanovic, Y.; Esuault, C.; Yang, T.; Jacquet, M.; Janin, J.; Duguet, M.; Perasso, R.; et al. Fine organization of Bombyx mori fibroin heavy chain gene. *Nucl. Acid Res.* **2000**, *28*, 2413–2419. [CrossRef]
15. Xu, G.; Gong, L.; Yang, Z.; Liu, X.Y. What makes spider silk fibers so strong? From molecular-crystallite network to hierarchical network structures. *Soft Matter* **2014**, *10*, 2116–2123. [CrossRef]
16. Murphy, A.R.; Kaplan, D.L. Biomedical applications of chemically-modified silk fibroin. *J. Mater. Chem.* **2009**, *19*, 6443–6450. [CrossRef]
17. Maziz, A.; Leprette, O.; Boyer, L.; Blatché, C.; Bergaud, C. Tuning the properties of silk fibroin biomaterial via chemical cross-linking. *Biomed. Phys. Eng. Express* **2018**, *4*, 065012. [CrossRef]
18. Bessonov, I.V.; Rochev, Y.A.; Arkhipova, A.Y.; Kopitsyna, M.N.; Bagrov, D.V.; Karpushkin, E.A.; Bibikova, T.N.; Moysenovich, A.M.; Soldatenko, A.S.; Nikishin, I.I.; et al. Fabrication of hydrogel scaffolds via photocrosslinking of methacrylated silk fibroin. *Biomed. Mater.* **2019**, *14*, 034102. [CrossRef]
19. Rockwood, D.N.; Preda, R.C.; Yücel, T.; Wang, X.; Lovett, M.L.; Kaplan, D.L. Materials fabrication from Bombyx mori silk fibroin. *Nat. Protoc.* **2011**, *6*, 1612–1631. [CrossRef]
20. Efremov, Y.M.; Bagrov, D.V.; Dubrovin, E.V.; Shaitan, K.V.; Yaminskii, I.V. Atomic force microscopy of animal cells: Advances and prospects. *Biophysics* **2011**, *56*, 257–267. [CrossRef]
21. Kim, D.H.; Wirtz, D. Focal adhesion size uniquely predicts cell migration. *FASEB J.* **2013**, *27*, 1351–1361. [CrossRef] [PubMed]

22. Hu, X.; Shmelev, K.; Sun, L.; Gil, E.S.; Park, S.H.; Cebe, P.; Kaplan, D.L. Regulation of silk material structure by temperature-controlled water vapor annealing. *Biomacromolecules* **2011**, *12*, 1686–1696. [CrossRef] [PubMed]
23. Byler, D.M.; Susi, H. Examination of the secondary structure of proteins by deconvolved FTIR spectra. *Biopolymers* **1986**, *25*, 469–487. [CrossRef]
24. Payne, K.J.; Veis, A. Fourier transform ir spectroscopy of collagen and gelatin solutions: Deconvolution of the amide I band for conformational studies. *Biopolymers* **1988**, *27*, 1749–1760. [CrossRef]
25. Meinel, L.; Hofmann, S.; Karageorgiou, V.; Kirker-head, C.; Mccool, J.; Gronowicz, G.; Zichner, L.; Langer, R.; Novakovic, G.V.; Kaplan, D.L. The inflammatory responses to silk films in vitro and in vivo. *Biomaterials* **2005**, *26*, 147–155. [CrossRef] [PubMed]
26. Karageorgiou, V.; Meinel, L.; Hofmann, S.; Malhotra, A.; Volloch, V.; Kaplan, D. Bone morphogenetic protein-2 decorated silk fibroin films induce osteogenic differentiation of human bone marrow stromal cells. *J. Biomed. Mater. Res. Part A* **2004**, *71*, 528–537. [CrossRef]
27. Mitropoulos, A.; Burpo, F.; Nguyen, C.; Nagelli, E.; Ryu, M.; Wang, J.; Sims, R.; Woronowicz, K.; Wickiser, J. Noble Metal Composite Porous Silk Fibroin Aerogel Fibers. *Materials* **2019**, *12*, 894. [CrossRef] [PubMed]
28. Ronca, A.; Maiullari, F.; Milan, M.; Pace, V.; Gloria, A.; Rizzi, R.; De Santis, R.; Ambrosio, L. Surface functionalization of acrylic based photocrosslinkable resin for 3D printing applications. *Bioact. Mater.* **2017**, *2*, 131–137. [CrossRef]
29. Keddie, J.L.; Jones, R.A.L.; Cory, R.A. Size-dependent depression of the glass transition temperature in polymer films. *EPL* **1994**, *27*, 59–64. [CrossRef]
30. Guo, Q. *Polymer Morphology: Principles, Characterization, and Processing*; John Wiley & Sons: Hoboken, NJ, USA, 2016; ISBN 978-1-118-45215-8.
31. Hu, X.; Kaplan, D.; Cebe, P. Determining beta-sheet crystallinity in fibrous proteins by thermal analysis and infrared spectroscopy. *Macromolecules* **2006**, *39*, 6161–6170. [CrossRef]
32. McGill, M.; Holland, G.P.; Kaplan, D.L. Experimental Methods for Characterizing the Secondary Structure and Thermal Properties of Silk Proteins. *Macromol. Rapid Commun.* **2019**, *40*, 1–14. [CrossRef] [PubMed]
33. Kim, U.-J.; Park, J.; Joo Kim, H.; Wada, M.; Kaplan, D.L. Three-dimensional aqueous-derived biomaterial scaffolds from silk fibroin. *Biomaterials* **2005**, *26*, 2775–2785. [CrossRef] [PubMed]
34. Li, C.; Hotz, B.; Ling, S.; Guo, J.; Haas, D.S.; Marelli, B.; Omenetto, F.; Lin, S.J.; Kaplan, D.L. Regenerated silk materials for functionalized silk orthopedic devices by mimicking natural processing. *Biomaterials* **2016**, *110*, 24–33. [CrossRef] [PubMed]
35. Melke, J.; Midha, S.; Ghosh, S.; Ito, K.; Hofmann, S. Silk fibroin as biomaterial for bone tissue engineering. *Acta Biomater.* **2016**, *31*, 1–16. [CrossRef]
36. Thai, T.H.; Nuntanaranont, T.; Kamolmatyakul, S.; Meesane, J. In Vivo evaluation of modified silk fibroin scaffolds with a mimicked microenvironment of fibronectin/decellularized pulp tissue for maxillofacial surgery. *Biomed. Mater.* **2017**, *13*, 015009. [CrossRef]
37. Perrone, G.S.; Leisk, G.G.; Lo, T.J.; Moreau, J.E.; Haas, D.S.; Papenburg, B.J.; Golden, E.B.; Partlow, B.P.; Fox, S.E.; Ibrahim, A.M.S.; et al. The use of silk-based devices for fracture fixation. *Nat. Commun.* **2014**, *5*, 3385. [CrossRef]
38. Midha, S.; Murab, S.; Ghosh, S. Osteogenic signaling on silk-based matrices. *Biomaterials* **2016**, *97*, 133–153. [CrossRef]
39. Kim, S.H.; Yeon, Y.K.; Lee, J.M.; Chao, J.R.; Lee, Y.J.; Seo, Y.B.; Sultan, M.T.; Lee, O.J.; Lee, J.S.; Yoon, S.I.L.; et al. Precisely printable and biocompatible silk fibroin bioink for digital light processing 3D printing. *Nat. Commun.* **2018**, *9*, 1–14.
40. Hytönen, V.P.; Wehrle-Haller, B. Protein conformation as a regulator of cell-matrix adhesion. *Phys. Chem. Chem. Phys.* **2014**, *16*, 6342–6357. [CrossRef]
41. Pacifici, A.; Laino, L.; Gargari, M.; Guzzo, F.; Luz, A.V.; Polimeni, A.; Pacifici, L. Decellularized hydrogels in bone tissue engineering: A topical review. *Int. J. Med. Sci.* **2018**, *15*, 492–497. [CrossRef]
42. Xie, J.; Zhang, D.; Zhou, C.; Yuan, Q.; Ye, L.; Zhou, X. Substrate elasticity regulates adipose-derived stromal cell differentiation towards osteogenesis and adipogenesis through β-catenin transduction. *Acta Biomater.* **2018**, *79*, 83–95. [CrossRef] [PubMed]
43. McBeath, R.; Pirone, D.M.; Nelson, C.M.; Bhadriraju, K.; Chen, C.S. Cell Shape, Cytoskeletal Tension, and RhoA Regulate Stem Cell Lineage Commitment. *Dev. Cell* **2004**, *6*, 483–495. [CrossRef]

44. Sonowal, H.; Kumar, A.; Bhattacharyya, J.; Gogoi, P.; Jaganathan, B. Inhibition of actin polymerization decreases osteogeneic differentiation of mesenchymal stem cells through p38 MAPK pathway. *J. Biomed. Sci.* **2013**, *20*, 71. [CrossRef] [PubMed]
45. Goldmann, W.H.; Auernheimer, V.; Thievessen, I.; Fabry, B. Vinculin, cell mechanics and tumour cell invasion. *Cell Biol. Int.* **2013**, *37*, 397–405. [CrossRef] [PubMed]
46. Bays, J.L.; DeMali, K.A. Vinculin in cell-cell and cell-matrix adhesions. *Cell. Mol. Life Sci.* **2017**, *74*, 2999–3009. [CrossRef] [PubMed]
47. Jannatbabaei, A.; Tafazzoli-Shadpour, M.; Seyedjafari, E.; Fatouraee, N. Cytoskeletal remodeling induced by substrate rigidity regulates rheological behaviors in endothelial cells. *J. Biomed. Mater. Res. Part A* **2019**, *107*, 71–80. [CrossRef]
48. Matsuoka, F.; Takeuchi, I.; Agata, H.; Kagami, H.; Shiono, H.; Kiyota, Y.; Honda, H.; Kato, R. Morphology-Based Prediction of Osteogenic Differentiation Potential of Human Mesenchymal Stem Cells. *PLoS ONE* **2013**, *8*, e55082. [CrossRef]
49. Keselowsky, B.G.; Collard, D.M.; García, A.J. Surface chemistry modulates focal adhesion composition and signaling through changes in integrin binding. *Biomaterials* **2004**, *25*, 5947–5954. [CrossRef]
50. Keselowsky, B.G.; Collard, D.M.; García, A.J. Integrin binding specificity regulates biomaterial surface chemistry effects on cell differentiation. *Proc. Natl. Acad. Sci. USA* **2005**, *102*, 5953–5957. [CrossRef] [PubMed]
51. Motta, A.; Migliaresi, C.; Lloyd, A.W.; Denyer, S.P.; Santin, M. Serum protein absorption on silk fibroin fibers and films: Surface opsonization and binding strength. *J. Bioact. Compat. Polym.* **2002**, *17*, 23–35. [CrossRef]

© 2020 by the authors. Licensee MDPI, Basel, Switzerland. This article is an open access article distributed under the terms and conditions of the Creative Commons Attribution (CC BY) license (http://creativecommons.org/licenses/by/4.0/).

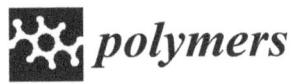

Article

Thermosensitive Chitosan–Gelatin–Glycerol Phosphate Hydrogels as Collagenase Carrier for Tendon–Bone Healing in a Rabbit Model

Yu-Min Huang [1,2,3], Yi-Cheng Lin [2,3], Chih-Yu Chen [2,3], Yueh-Ying Hsieh [2,3], Chen-Kun Liaw [2,3], Shu-Wei Huang [1], Yang-Hwei Tsuang [2,3], Chih-Hwa Chen [4] and Feng-Huei Lin [1,5,*]

1. Department of Biomedical Engineering, National Taiwan University, Taipei 100, Taiwan; yellowcorn0326@yahoo.com.hk (Y.-M.H.); judyya1022@gmail.com (S.-W.H.)
2. Department of Orthopedics, Shuang Ho Hospital, Taipei Medical University, Taipei 100, Taiwan; iam4290@gmail.com (Y.-C.L.); aleckc2424@gmail.com (C.-Y.C.); ianhsie@gmail.com (Y.-Y.H.); d92008@yahoo.com.tw (C.-K.L.); tsuangyh@gmail.com (Y.-H.T.)
3. Department of Orthopedics, School of Medicine, College of Medicine, Taipei Medical University, Taipei 100, Taiwan
4. Department of Orthopedics, Taipei Medical University – Shuang Ho Hospital, School of Medicine, College of Medicine, School of Biomedical Engineering, College of Biomedical Engineering, Research Center of Biomedical Device, Taipei Medical University, Taipei 100, Taiwan; chihhwachen@gmail.com
5. Institute of Biomedical Engineering & Nanomedicine, National Health Research Institutes, Miaoli County 360, Taiwan
* Correspondence: double@ntu.edu.tw; Tel.: +886-2-2732-0443

Received: 26 December 2019; Accepted: 8 February 2020; Published: 13 February 2020

Abstract: Healing of an anterior cruciate ligament graft in bone tunnel yields weaker fibrous scar tissue, which may prolong an already prolonged healing process within the tendon–bone interface. In this study, gelatin molecules were added to thermosensitive chitosan/β-glycerol phosphate disodium salt hydrogels to form chitosan/gelatin/β-glycerol phosphate (C/G/GP) hydrogels, which were applied to 0.1 mg/mL collagenase carrier in the tendon–bone junction. New Zealand white rabbit's long digital extensor tendon was detached and translated into a 2.5-mm diameter tibial plateau tunnel. Thirty-six rabbits underwent bilateral surgery and hydrogel injection treatment with and without collagenase. Histological analyses revealed early healing and more bone formation at the tendon–bone interface after collagenase partial digestion. The area of metachromasia significantly increased in both 4-week and 8-week groups after collagenase treatment ($p < 0.01$). Micro computed tomography showed a significant increase in total bone volume and bone volume/tissue volume in the 8 weeks after collagenase treatment, compared with the control group. Load-to-failure was significantly higher in the treated group at 8 weeks (23.8 ± 8.13 N vs 14.3 ± 3.9 N; $p = 0.008$). Treatment with collagenase digestion resulted in a 66% increase in pull-out strength. In conclusion, injection of C/G/GP hydrogel with collagenase improves tendon-to-bone healing in a rabbit model.

Keywords: hydrogel; tendon–bone healing; collagenase digestion; rabbit model

1. Introduction

The anterior cruciate ligament (ACL) is one of the most commonly injured soft tissue structures, with an injury incidence of one in 3000 in the American population [1]. Primary repair is not effective for treating injured ACL tissue due to poor healing potential. ACL reconstruction with autografts or allografts has a greater than 90% success rate in restoring knee stability or previous functional performance [2]. Autologous bone-patellar-tendon grafts (BPTB) achieve good results due to direct bone-to-bone healing in femoral and tibial tunnels. However, the harvesting of BPTB results in significant donor morbidity, anterior knee pain, and higher rates of osteoarthritis [3]. Autologous hamstring tendon grafts restore function to a similar extent as BPTB grafts, but have a higher graft failure rate and revision rate [4,5]. Currently, there is an unmet need for interventions that can enhance tendon-to-bone healing to attain a lower graft failure rate and better functional outcomes.

One of the keys to successful ACL reconstruction is healing of the tendon graft to bone. In general, there are transition zones between the tendon-to-bone interface, with four tissue types as follows: tendon, unmineralized fibrocartilage, mineralized fibrocartilage, and bone [6]. Rodeo et al. found a healing process with a fibrovascular interface between the tendon and bone in an extra-articular model [7]. The healing of the tendon-to-bone interface is slow due to the relative avascularity of the fibrocartilage zone and bone loss [8]. The histological arrangement is similar to using Sharpey fibers as indirect insertions, such as in the medial collateral ligament of the knee joint. Because of the scar tissue between the tendon and bone interface, there is a weak connection between tendon grafts and bones in ACL reconstruction, especially in the first six weeks [9]. Besides, tenocytes are surrounded in a dense extracellular matrix (ECM), which limits their mobility and proliferation. Our previous study showed that partial collagenase digestion could improve healing for osteochondral defects in a mice model by degrading the extracellular matrix [10]. Hong et al. also demonstrated the injection of hydrogel, which induced macrophages to produce MMP-9 enzyme for tissue repair by promoting extracellular matrix remodeling [11]. In this present study, we aimed to investigate the effects of low dose collagenase for partial digestion of the ECM in improving tendon–bone healing.

Thermosensitive polymers have gained great attention in drug delivery, cell encapsulation, and tissue engineering [12]. These polymers can be carriers for drug delivery through minimal invasive surgery. Injectable polymers have several advantages in drug delivery, including easy preparation, prolonged and localized drug release, and low systemic toxicity [13]. Chitosan/glycerol phosphate (GP; disodium salt) is one of the widely exploited thermosensitive hydrogels and has drug-delivery potential [14]. Chitosan is an aminopolysaccharide derived from alkaline deacetylation of chitin, which is present in fungal cell walls. In combination with glycerol phosphate, the chitosan/GP hydrogel becomes thermosensitive in diluted acids and can proceed gelation around body temperature [15]. Chitosan/GP hydrogel can be in a liquid state at physiological PH and then a gel when heated at 37 °C. Ahmadi et al. indicated chitosan-GP is a biocompatible hydrogel for mesenchymal stem cell proliferation at certain concentrations [16]. Roughley et al. demonstrated that lack for firm structure of chitosan/GP hydrogel may not be suitable for cell culture [17]. In order to improve the gelation strength, gelatin was added to the thermosensitive chitosan-GP hydrogels. Gelatin is a biocompatible and biodegradable polymer for tissue engineering [18]. Cheng et al. showed chitosan–gelatin–glycerol phosphate hydrogels are biocompatible for nucleus pulposus cell regeneration with better gel strength [19]. When gelatin is added to the chitosan/GP solution at low temperature, hydrogen bonds exist between the OH group of gelatin and the OH and NH_2 groups of chitosan. Due to high hydrophilicity of gelatin, the binding between gelatin and water can decrease the mobility of chitosan molecules.

The purpose of this study was to determine whether partial collagenase digestion can improve tendon-to-bone healing in a bone tunnel animal model. We hypothesized that the administration of a collagenase hydrogel could degrade the fibrous tissue and lead to the early regeneration of fibrovascular tissue during the healing process. Furthermore, we hypothesized that improved histological results would indicate stronger tendon-to-bone fixation.

2. Materials and Methods

2.1. Collagenase Hydrogel Preparation

A sterile 0.1% collagenase solution was prepared by mixing 10 mg of Liberase TM Research Grade (Roche, Mannheim, Germany) with 10 mL of Dulbecco's modified Eagle's medium (DMEM). An injectable and thermosensitive chitosan/gelatin/glycerol phosphate (C/G/GP) hydrogel was used as a controlled release system for enzyme delivery [19]. For hydrogel preparation, 2.5% chitosan (degree of deacetylation >95%, Kiotek, Taiwan) and 1% gelatin (G1890, Sigma-Aldrich, St. Louis, MO, USA) were dissolved in 0.1 M acetic acid (242,853, Sigma-Aldrich, St. Louis, MO, USA). Glycerol 2-phosphate disodium salt hydrate (β-GP, G6251, Sigma-Aldrich, St. Louis, MO, USA) was dissolved in water and filtered using a 0.22-μm filter (Millex-GV; Millipore, Billerica, MA, USA) for sterilization. The C/G/GP solution was adjusted to pH 7.4, which was suitable for collagenase release. The thermosensitive C/G/GP hydrogel gelled at 37 °C and was injectable with the collagenase preparation.

2.2. Cytotoxicity Evaluation

The Achilles tendon was harvested from New Zealand White rabbit following euthanasia. Tendon was washed by PBS buffer (Biochrom, Berlin, Germany) with 100 U/mL penicillin (P4083; Sigma, St. Louis, MO, USA) and cut into small pieces. Minced pieces were placed in a 48-well culture plate and incubated with 0.2% (*v/v*) collagenase (C0130; Sigma, St. Louis, MO, USA) for 30 min. Further DMEM containing 10% fetal bovine serum (Cat. No. 100-106; Gemini Bio-products, West Sacramento, CA, USA) was added and cultured. Tenocytes of passage less than 5 were used in the cytotoxicity test. Tenocytes were seeded in the 96-well cell culture plates (92,096; TPP, Ho Chi Minh City, Vietnam) at a density of 10,000 cells/well and cultured in DMEM mixture F-12 ham medium (DMEM-F12, D8900; Sigma). After incubation for 18 h, the cells were washed with PBS buffer and the solution of hydrogel was added into the culture well (200 μL/well). WST-1 (Cell Proliferation Reagent WST-1; Roche, Mannheim, Germany) was measured at day 1 and 3 to check the cell viability. OD values was measured at 450 nm by using enzyme-linked immune-sorbent assay (ELISA) reader (Sunrise Remote; Tecan, Durham, NC, USA). LDH (CytoTox96 Non-Radioactive Cytotoxicity Assay; Promega, Madison, WI, USA) assay was applied to identify the collagenase C/G/GP hydrogel toxicity to the tenocyte. LDH was measured with a 30 min coupled enzymatic assay and measured using ELISA reader at a wavelength of 490 nm.

2.3. Release Profile of Collagenase from C/G/GP Hydrogel

In order to obtain the release profile of collagenase, 1 cc C/G/GP hydrogel was prepared with 0.1 mg/mL collagenase solution. Once the hydrogel gelled completely, the material was immersed in a 24-well plate with 1.5 mL PBS buffer solution/well at 37 °C. The hydrogel soaked in the PBS solution was collected at regular time intervals (30 min, and 1, 3, 8, 24, 48, and 96 h) and refilled with fresh PBS. The PBS obtained was measured by UV-vis spectrometer analysis at 258 nm to quantify the release profile of collagenase from the C/G/GP hydrogel.

2.4. Animal Study Design

Thirty-six male New Zealand white rabbits (weighing 3.25 ± 0.15 kg) (from BioLASCO Taiwan Co., Taipei, Taiwan) were used in this study. All the rabbits were randomly allocated to subgroup by a random number generator. The collagenase-containing hydrogel was injected into the right knee and the left knee was injected with C/G/GP hydrogel only without collagenase. Eighteen rabbits were prepared for histological and image evaluation at 4 and 8 weeks, respectively (Figure 1). Four rabbits per group were prepared for histological study, and six rabbits per group were evaluated by microcomputed tomography (micro-CT) analysis. Eight rabbits from each group were subjected to a biomechanical pull-out test.

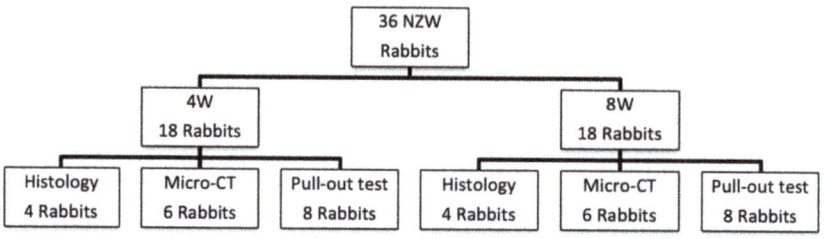

Figure 1. Animal study design.

Flowchart shows the distribution in our animal study.

2.5. Surgical Procedure

All procedures were carried out according to the Guide for the Care and Use of Laboratory Animals and approved by Taipei Medical University (LAC-101-0210) on 7th March 2013. All rabbits received a preoperative dose of intramuscular cefazolin sodium (0.1 mg/kg) as a prophylactic antibiotic. Ketamine (40 mg/kg) with xylazine (5 mg/kg) (Rompun; Bayer Healthcare, Leverkusen, Germany) was intramuscularly injected to induce general anesthesia. Using an animal model similar to that developed by Chang et al., [20] the long digital extensor tendon was dissected from the lateral femoral condyle. We created a 2.5-mm-diameter tunnel by electrical drill at the proximal tibial metaphysis at a 30-degree angle from lateral to medial relative to the long bone axis. In our study, the bony tunnel was extra-articular without connection to the knee joint. The long digital extensor tendon was detached from the origin site and translated into the drill hole and fixed with a 3-0 vicryl suture at the medial capsule of the knee joint. Then, 0.5 mL C/G/GP hydrogel with 0.1 mg/mL Liberase TM Research Grade was injected into the bone tunnel through a 24-gauge needle after tendon passed through on right knee. The left knee underwent the same procedure with hydrogel injection without collagenase. The rabbits were allowed to freely exercise in the same cage without restrictions. Euthanasia has been performed under CO2 chamber at 4 and 8 weeks, and the animals were randomly prepared for histological and biomechanical examination.

2.6. Histological Examination

The specimens were fixed with 10% formalin for 24 h and decalcified in a graded series of alcohol. The proximal tibia was embedded in paraffin and sliced into 5-μm-thick sections sagittal to the bone tunnel. The samples were sectioned parallel to the longitudinal axis of the tibial bone tunnel. Staining was performed using hematoxylin and eosin (H&E) for the identification of collagenous fiber tissue. Histological sections were examined using a light microscope. Healing between the tendon and the bone tunnel was assessed by identifying new tissue (woven bone formation or fibrovascular granulation tissue).

The area of fibrovascular tissue at the bone-tendon interface was identified as the area of metachromasia. The total area of metachromasia was measured for all histological specimens using Image J imaging software. The area of metachromasia was gauged by two independent observers for histological measurements.

2.7. Micro-Computed Tomography Evaluation

At 4 and 8 weeks after surgery, six rabbits from each group were sacrificed. The specimens were prepared for micro-CT analysis (Skyscan 1176 μCT System; Bruker, Kontich, Belgium). A 9-μm resolution along the long axis of the tibial bone tunnel was acquired using a consecutive micro-CT imager (50 kV X-ray voltage and 200 μA electric current). The scans yielded reconstructed 3D data sets with a voxel size of 9 μm, which were evaluated using a CT automated image analysis system (Bruker, Kontich, Belgium). The volume of interest (VOI) contained the whole bone tunnel with the tendon graft and the surroundings. Total bone volume (TBV, mm^3) and the bone volume fraction (BV/TV) were measured from the number of bone voxels and the total voxel number in the VOI. Trabecular thickness (TbTh, μm) was measured from the trabecular architecture around the tendon graft. Bone mineral density (BMD, g/mm^3) was also calculated using the analysis system. The two investigators who evaluated the CT data were blinded to the animal study.

2.8. Biomechanical Testing

Eight rabbits from each group were selected for biomechanical testing.

Before testing, we carefully dissected the tendon from the distal end of the insertion and removed all the other soft tissue. We used a Bose ElectroForce 3510 Fatigue Tester at an elongation rate of 10 mm/min. The graft was secured on the clamps 3 cm away from the lateral tibial aperture, allowing tensile loading along the long axis of the tibial bone tunnel. Following a static preload of 1 N for 2 min, ultimate load-to-failure in uniaxial tension at 20 mm per minute was determined from the load-deformation curve. The load-deformation curve was recorded, from which the ultimate load-to-failure and stress (load/cross-section area) of the tendon graft were recorded.

2.9. Statistical Analyses

Before the animal study, a power analysis was performed to determine the number of animals needed to prevent a type II error. In a previous study, Gulotta et al. showed that the TBV determined by μCT imaging was 27.0 ± 8.1 mm^3 for experimental tibia [20]. Five specimens per group provided a power of 0.80 with $\alpha = 0.05$. For biomechanical testing in this study, we performed a pilot study at our institution that showed that the ultimate pull-out strength was 13.1 ± 8.4 N". This analysis showed that 16 specimens (8 specimens in each group) could achieve a power of 0.80 with $\alpha = 0.05$. The obtained biomechanical data and CT analysis were compared using paired t-tests. All statistical analyses were performed using SPSS 18.0 software (SPSS Inc., Chicago, IL, USA). $p < 0.05$ was considered statistically significant.

3. Results

3.1. Cytotoxicity

Figure 2a shows the results of WST-1 assay of the C/G/GP hydrogel with collagenase at day 1 and 3. There is no significant difference between the C/G/GP hydrogels with or without collagenase, compared with the monolayer cultured group (n = 5, $p > 0.05$). Figure 2b shows the LDH assay among control medium and C/G/GP hydrogels groups. The results of LDH assay show no significant difference among experimental agents (n = 5, $p > 0.05$). The results of LDH assay illustrate the C/G/GP hydrogel with collagenase has no cytotoxic effect on tenocyte. Hence, we choose the C/G/GP hydrogel with 0.1% collagenase for our experiments according to the safety test.

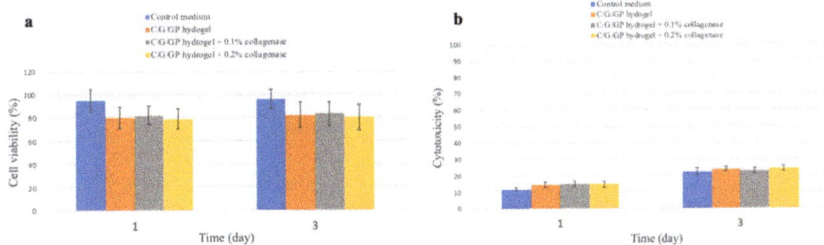

Figure 2. Cytotoxic of chitosan/gelatin/β-glycerol phosphate (C/G/GP) with collagenase to tenocyte. Figure 2a shows water soluble tetrazolium salt-1 (WST-1) assay (n = 5, $p > 0.05$) under different collagenase concentration and Figure 2b shows lactate dehydrogenase assay (n = 5, $p > 0.05$).

3.2. Collagenase Release Profile

The collagenase release profile from the C/G/GP hydrogel is given in Figure 3. The release profile was observed at the quick release in the first 30 min, and at sustained release until 4 days. Fifty-one percent collagenase was released from hydrogel in the first 30 min, and saturation was continued after 48 h. The profile illustrates that collagenase could release faster and reach higher cumulative release rate (92.5%).

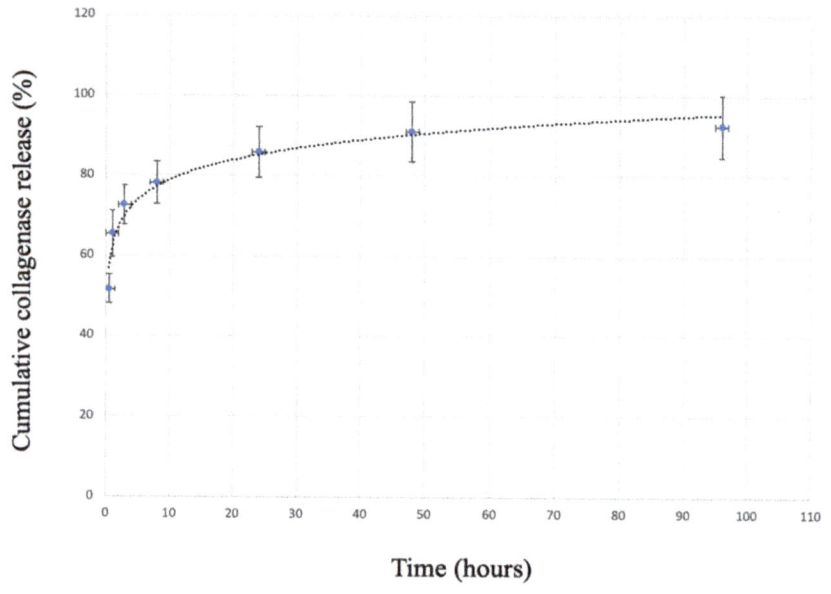

Figure 3. Release profile of the collagenases loaded in C/G/GP hydrogel. The release profile was evaluated by incubation C/G/GP hydrogel in PBS with collagenase concentration of 0.1 mg/mL. The error bars designed the standard deviation.

3.3. Gross Observations

After necropsy, all specimens were examined from the knee joint to the middle aspect of the tibia. The knee joint contained clear synovial fluid without signs of infection. Synovitis with hypertrophy of the synovium was not observed in this study. Gross observation revealed no evidence of an adverse effect after using collagenase digestion in the knee joint or surroundings. Soft tissue around the bone tunnel at the tibial metaphysis showed no obvious necrosis or defects in either group. The bony defect in the bone tunnel was clearly identified in the control group (Figure 4a,b). In the control group, only one specimen was found to have bone formation in the bone orifice at 4 weeks, and six specimens were observed to have bone formation at 8 weeks. Bone tunnel healing with new bone formation was observed in all study groups, especially in the 8-week groups (Figure 4c,d). All surgical wounds healed smoothly, without wound dehiscence or infection.

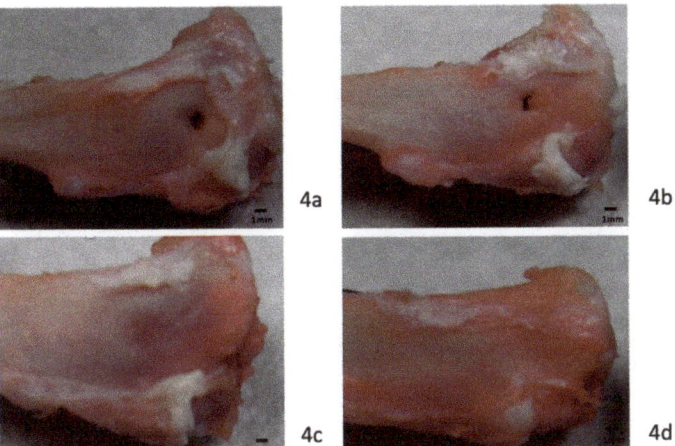

Figure 4. Necropsy examination. Photography at necropsy showed bone-tunnel healing in a rabbit model. A bony defect was identified in the control group at 4 weeks (**a**) and 8 weeks (**b**). Better bony ingrowth with a decrease in tunnel diameter was seen in the study group at 4 weeks (**c**) and 8 weeks (**d**) (scale bar: 1 mm).

3.4. Histology

3.4.1. Four-Week Specimens

Healing at the tendon–bone interface was observed with the formation of fibrovascular tissue in the treated group (Figure 5a,b). A large gap between the tendon and bone was clearly identified in the control group. More dense fibrous tissue formation was observed at both ends of the bone tunnel, especially in the medial orifice. There was very little bone formation and cartilage observed around the bone tunnel interface in the control group specimens.

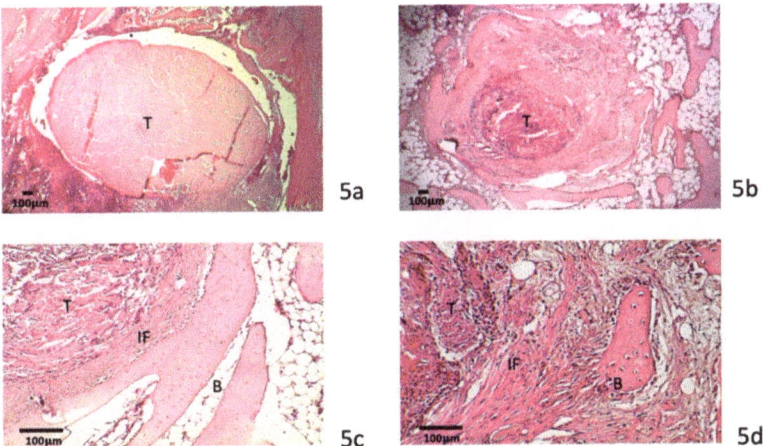

Figure 5. Four-week specimens histology. Histology of a tendon (T)-to-bone (B) interface (IF) in a 4-week specimen in the control group (**a**) and the study group (**b**). A wider interface with dense fibrovascular tissue formation in the study group was found. Sharpey fiber-like tissue was more observed 4 weeks after collagenase treatment (**c**,**d**) (scale bar: 100 μm).

In the collagenase digestion group, more maturation of the fibrous tissue was observed at the tendon–bone interface (Figure 5c,d). The morphology and structure of the tendon graft showed no significant difference from the control group. There was no evidence of a foreign body giant cell response in the treated group. Fibrovascular tissue at the tendon–bone interface was more mature in the collagenase-treated group than in the control group. Sharpey fiber-like tissue was also identified at tendon–bone interface. Incorporation of the tendon with collagen fibers and bone was more prominent in the treated specimens than in the control specimens. Proliferation of fibrous tissue at the tendon–bone interface was more advanced and apparent in the treated group than in the control group.

3.4.2. Eight-Week Specimens

At 8 weeks, there was more mature fibrovascular tissue incorporated at the tendon–bone interface after collagenase treatment. More Sharpey fiber-like tissue was observed in the treated specimens than the 4-week specimens. More bone formation with incorporation into the tendon graft was observed in the treated group than in the control group (Figure 6a,b). The tendon graft volume was decreased with early woven bone integration (Figure 6c,d). All treated specimens showed aggressive bone ingrowth at both ends of the bone tunnels. There was no significant difference in bone ingrowth between the medial and lateral orifice. The treated specimens had a wide interface zone and more woven bone formation than the control group.

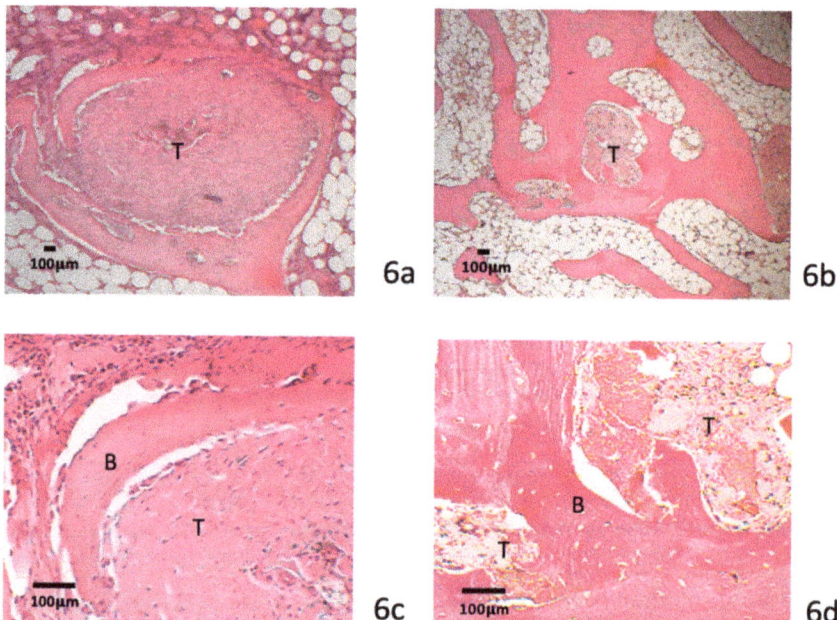

Figure 6. Eight-week specimens histology. Wider dense interface tissue between the tendon (T) and bone (B) in the 8-week study group (**a**,**b**). New bone formation with incorporation into the tendon graft is illustrated (**c**,**d**). The tendon graft was replaced by new bone formation after collagenase treatment. (scale bar: 100 μm).

3.4.3. Metachromasia

Histomorphometric analysis showed a significantly greater area of fibrovascular tissue at the tendon–bone interface in the collagenase group (Figure 7). Early and significant fibrous healing was demonstrated in the 4-week group ($p = 0.0008$). Significant increase of metachromacia was also demonstrated in 8-week group after collagenase treatment comparing with control group ($p = 0.004$).

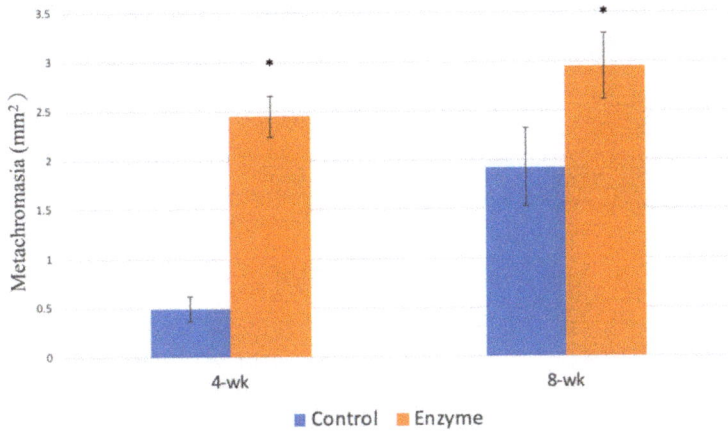

Figure 7. Metachromacia. The area of metachromcia was significantly increased in both 4-week and 8-week groups after collagenase treatment ($p < 0.01$). The error bars designed the standard deviation.

3.5. Micro-Computed Tomography Evaluation

New bone formation was evaluated by micro-CT, and the TBV was quantified. 3D reconstruction images showed more new bone formation in the treated specimens than in the control specimens at 8 weeks. In the control group, the bone tunnel with tendon graft was clearly identified in the CT images (Figure 8a,b). New bone formation with bone tunnel elimination was identified after collagenase treatment (Figure 8c,d). The bony tunnel area was decreased in the sagittal view after collagenase treatment. The new bone formation was prominent near the medial aperture in the coronal view. At 8 weeks, the TBV was significantly greater ($p = 0.003$) in the collagenase group (48.7 ± 6.8 mm^3) than in the control group (33 ± 6.6 mm^3) (Figure 9a). However, there were no differences in the TBV between the two groups at the 4-week timepoint. The percentage of new bone formation in the tendon–bone junction was analyzed as bone volume/tissue volume (BV/TV). We also noted the BV/TV was significantly greater ($p = 0.01$) in the collagenase group ($69.5 \pm 9.8\%$), compared with control group ($50.8 \pm 10.1\%$) at 8-week group. (Figure 9b). In addition, there were no significant differences in the trabecular thickness and BMD between the two groups.

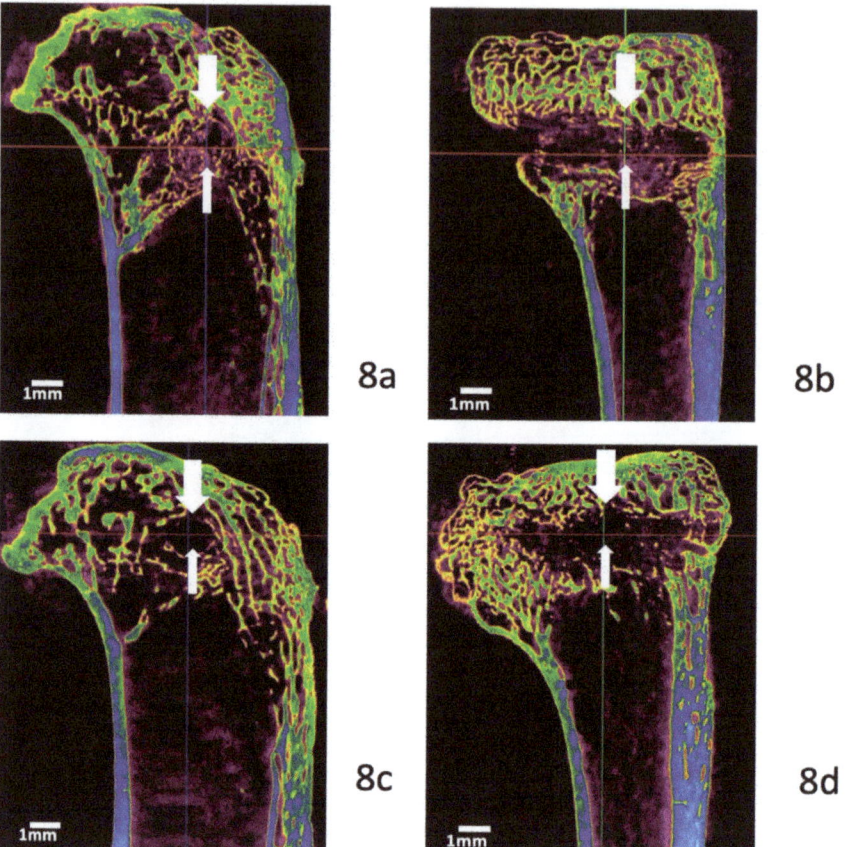

Figure 8. Microcomputed tomography (Micro-CT) examination showed bone formation. Micro-CT shows tendon graft and bone tunnel in an 8-week specimen from the control group. (**a,b**) Bone tunnel obliteration with new bone ingrowth was identified in sagittal and coronal views after enzyme treatment (**c,d**). Prominent bone ingrowth was observed in both tunnel orifices. Line arrow marks the tendon graft, and block arrows indicate the bone tunnel. (scale bar: 1 mm).

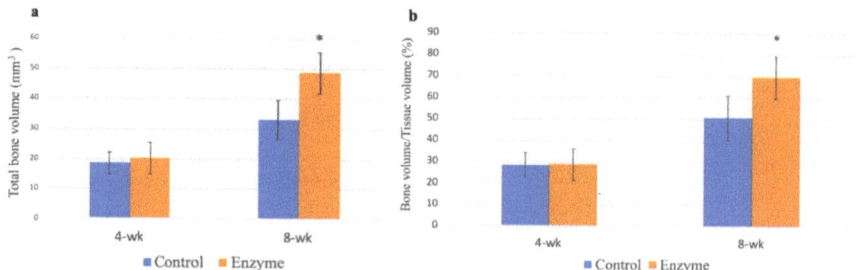

Figure 9. Micro-CT evaluation. Total bone volume (mm^3 ± SD) determined by micro-CT. The region of interest was measured following the long axis of the bone tunnel. (**a**) There was significantly more new bone formation at 8 weeks after collagenase treatment. Total bone volume/tissue volume was illustrated, and the ratio was significant 8 weeks after enzyme treatment. (**b**) $p < 0.05$, n = 6. The error bars designed the standard deviation.

3.6. Biomechanical Testing

Figure 10a demonstrated the original load-deformation curve in the 8-week enzyme-treated specimen. The ultimate load-to-failure and stress in the treated specimens were greater than that in the control group at 8 weeks. The mean maximal load at failure was 14.3 ± 3.9 N for the control group and 23.8 ± 8.13 N for the collagenase-treated group ($p = 0.008$). The stress was 4.65 ± 0.8 N/mm^2 for the control group comparing with 5.93 ± 1.02 N/mm^2 after collagenase treatment. ($p = 0.03$) Treatment with partial collagenase digestion resulted in an increased load of 66%. Most specimens failed in the bony tunnel, and only two specimens (2/8) failed near the proximal orifice in the collagenase group at 8 weeks. There were no differences between the two groups at 4 weeks.

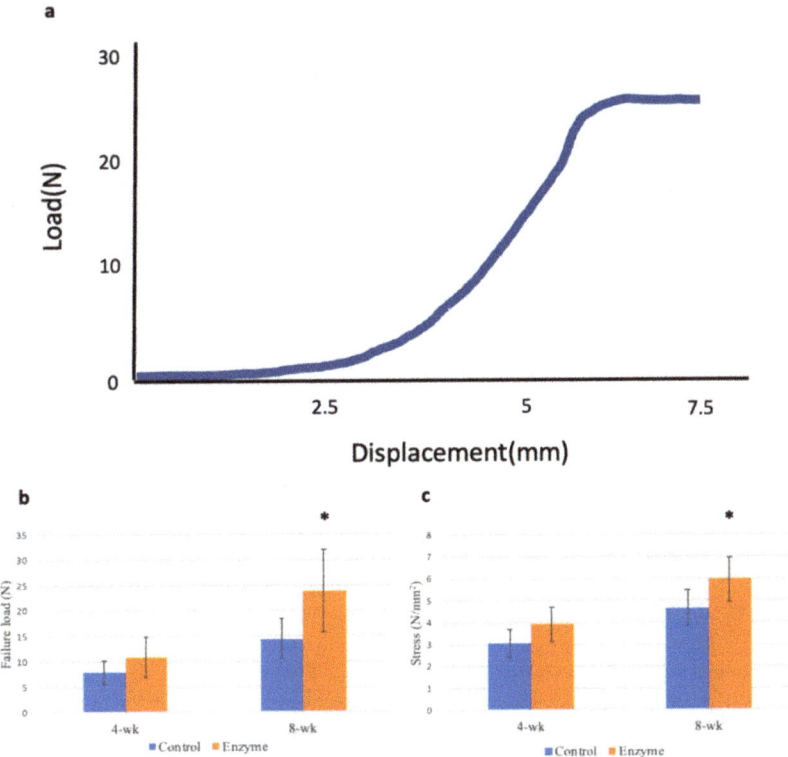

Figure 10. Biomechanical assessment. Original load-deformation curve in the 8-week enzyme-treated specimen. (**a**) Significantly higher tendon ultimate load-to-failure and stress were found in the enzyme treatment group than in the control group at 8 weeks. * $p < 0.05$, n = 8. (**b,c**) The error bars designed the standard deviation.

4. Discussion

Clinical outcomes after ACL reconstruction using soft tissue grafts are heavily dependent on secure integration at the tendon–bone interface. However, healing of a tendon graft in the bone tunnel involves fibrous scar tissue, which is mechanically inferior to a normal tendon insertion [21,22]. Bedi et al. reported differences in the graft healing process between the intra-articular and extra-articular ends of a bone tunnel [23]. Smith et al. showed more mature tendon-to-bone healing process can occur through a similar four-zone of integration in soft tissue grafts in the canine model [24]. The interface between the tendon and the bone tunnel is the weakest area for six weeks immediately after ACL reconstruction [9]. For these reasons, there is substantial interest in evaluating the effects of different agents on tendon–bone healing. Several experimental agents have been investigated, including recombinant human bone morphogenetic protein-2 (BMP-2) [7], transforming growth factor-ß1 (TGF-ß1) [25], magnesium-based bone adhesive [20], demineralized bone matrix [26], and calcium phosphate-hybridized tendon grafts [27]. Rodeo et al. demonstrated BMP-2 application can enhance bone ingrowth at the tendon–bone interface in a dog model [7]. Yamazaki et al. showed exogenous administration of TGF-β1 increased the bonding strength of the graft to the tunnel [25], and Mutsuzaki et al. developed a goat animal model using calcium phosphate hybridization with tendon graft to improve tendon–bone healing [27]. Yoon et al. found local PTH administration with an absorbable scaffold improved the biomechanical and histological outcomes in a rat model [28].

In this study, we evaluated the effect of partial collagenase digestion on tendon–bone interface healing. In the histological examination, we found more new bone formation and early perpendicular collagen fiber integration in the treated group. The appearance of new bone formation indicated a strengthening of the attachment at the tendon–bone interface, and this was confirmed by the mechanical pull-out test. By 8 weeks, the ultimate load-to-failure increased by approximately 66% in the treated group compared to that in the control group. We also used micro-CT to objectively determine the amount of new bone formation. New bone formation found at both ends of the bone tunnel indicated the effect of partial collagenase digestion.

The collagenase used in this study was Liberase TM Research Grade, which contains purified collagenase isoforms I and II and a neutral protease. Native collagenases used for tissue remodeling in humans are matrix metalloproteinases (MMPs). MMPs are a family of zinc-dependent endopeptidases that degrade the extracellular matrix between cells and adjacent tissues [29]. MMPs and their endogenous inhibitors play important roles in maintaining the integrity of the extracellular matrix [30]. The loss of balance between MMPs and their endogenous inhibitors results in degenerative tendinopathy or tendon rupture [31–35]. Bedi et al. reported that local delivery of MMP inhibitors improves tendon–bone healing after rotator cuff repair in a rabbit animal model [36]. In ACL rupture cases, the MMP-1 gene expression level is the highest in the synovium, and the highest level of MMP-13 gene expression is found in the ACL [37]. Demirag et al. reported the enhancement of tendon-to-bone healing by inhibiting MMPs in an ACL reconstruction rabbit model [38]. This suggests that both anabolic and catabolic factors are important in the healing process after ACL reconstruction.

There are two possible mechanisms by which partial collagenase digestion benefits healing. First, the tendon–bone interface is mainly filled with fibrous scar tissue. This fibrous scar tissue is weak, and there is poorer biomechanical resistance than in normal tendon–bone healing. Therefore, collagenase injection into the bone tunnel may digest the fibrous scar tissue and improve native woven bone formation. This was shown in the histology examination of the treated group in both 4-weeks and 8-weeks group. The second mechanism that may improve healing is woven bone overgrowth. In our study, early woven bone was observed in the CT images and histology examination of the treated group. Paiva et al. reviewed the pivotal functions of MMPs during development and bone regeneration in a knockout mouse model [39]. Bone remodeling is mediated by osteoclast recruitment, differentiation, and bone matrix resorption, followed by the recruitment and deposition of osteoblasts for new bone formation [40]. We hypothesized that the injection of a collagenase hydrogel alters the homeostasis of bone remodeling for early woven bone formation.

We acknowledge several limitations in our study. First, we used Liberase TM Research Grade as our experimental agent. Collagenase has been used to induce tendon degeneration or to release tenocytes in a previous study [41]. We used a C/G/GP hydrogel as a collagenase release system. The concentration of collagenase was evaluated in our laboratory, but the ideal amount of collagenase remains uncertain. Second, we conducted a power analysis to determine the numbers of animals required to minimize the concern of a type II error. Significant differences were identified in the 8-week groups, but CT parameters and biomechanical results showed no significant differences at 4 weeks. Therefore, type II errors could have biased this study. Finally, this was a preliminary study to evaluate the effect of partial collagenase digestion on tendon–bone healing. This study was an extra-articular graft model, which was different from ACL reconstruction. A positive effect could not be confirmed for ACL reconstruction surgery. Therefore, further animal studies using this model for ACL reconstruction are warranted.

5. Conclusions

Chitosan/gelatin/β-glycerol phosphate (C/G/GP) hydrogels was proposed as an ideal carrier for collagenase application to improve tendon–bone healing in a rabbit animal. To achieve a rapid tendon–bone healing, collagenase was incorporated with C/G/GP hydrogel could induce early and more woven bone formation. Biomechanical test revealed higher ultimate load-to-failure and stress after collagenase treatment. Histological examination and micro-CT observed more bone formation and perpendicular collagen fibers after the enzyme treatment. Our findings suggest a simple method and could be used to augment tendon-to-bone healing and call for further research.

Author Contributions: Investigation and writing: Y.-M.H., Y.-C.L., and C.-Y.C. Project administration: Y.-Y.H., S.-W.H. Resources and software: C.-K.L., S.-W.H. Results evaluation/Supervision: Y.-H.T., C.-H.C., and F.-H.L. All authors have read and agreed to the published version of the manuscript.

Funding: This research was funded by The Ministry of Science and Technology (102-2314-B-038-009). The sponsor had no role in the study design, data collection, data analysis, or manuscript writing.

Acknowledgments: The authors would like to thank Jui-Sheng Sun for his expert assistance with this project in National Taiwan University Hospital.

Conflicts of Interest: The authors declare no conflict of interest.

References

1. Frank, C.B.; Jackson, D.W. Current concepts review—The science of reconstruction of the anterior cruciate ligament. *J. Bone Jt. Surg.* **1997**, *79*, 1556–1576. [CrossRef] [PubMed]
2. Freedman, K.B.; D'Amato, M.J.; Nedeff, D.D.; Kaz, A.; Bach, B.R. Arthroscopic anterior cruciate ligament reconstruction: A metaanalysis comparing patellar tendon and hamstring tendon autografts. *Am. J. Sports Med.* **2003**, *31*, 2–11. [CrossRef] [PubMed]
3. Poehling-Monaghan, K.L.; Salem, H.; Ross, K.E.; Secrist, E.; Ciccotti, M.C.; Tjoumakaris, F.; Ciccotti, M.G.; Freedman, K.B. Long-term outcomes in anterior cruciate ligament reconstruction: A systematic review of patellar tendon versus hamstring autografts. *Orthop. J. Sports Med.* **2017**, *5*. [CrossRef]
4. Gifstad, T.; Foss, O.A.; Engebretsen, L.; Lind, M.; Forssblad, M.; Albrektsen, G.; Drogset, J.O. Lower risk of revision with patellar tendon autografts compared with hamstring autografts: A registry study based on 45,998 primary ACL reconstructions in Scandinavia. *Am. J. Sports Med.* **2014**, *42*, 2319–2328. [CrossRef]
5. Samuelsen, B.T.; Webster, K.E.; Johnson, N.R.; Hewett, T.E.; Krych, A.J. Hamstring Autograft versus Patellar Tendon Autograft for ACL Reconstruction: Is There a Difference in Graft Failure Rate? A Meta-analysis of 47,613 Patients. *Clin. Orthop. Relat. Res.* **2017**, *475*, 2459–2468. [CrossRef]
6. Cooper, R.R.; Misol, S. Tendon and ligament insertion. A light and electron microscopic study. *J. Bone Jt. Surg.* **1970**, *52*, 1–20. [CrossRef]
7. Rodeo, S.A.; Suzuki, K.; Deng, X.-H.; Wozney, J.; Warren, R.F. Use of Recombinant Human Bone Morphogenetic Protein-2 to Enhance Tendon Healing in a Bone Tunnel. *Am. J. Sports Med.* **1999**, *27*, 476–488. [CrossRef]
8. Wong, M.W.N.; Qin, L.; Tai, J.K.O.; Lee, S.K.M.; Leung, K.S.; Chan, K.M. Engineered allogeneic chondrocyte pellet for reconstruction of fibrocartilage zone at bone-tendon junction? A preliminary histological observation. *J. Biomed. Mater. Res.* **2004**, *70B*, 362–367. [CrossRef]
9. Tomita, F.; Yasuda, K.; Mikami, S.; Sakai, T.; Yamazaki, S.; Tohyama, H. Comparisons of intraosseous graft healing between the doubled flexor tendon graft and the bone-patellar tendon-bone graft in anterior cruciate ligament reconstruction. *Arthroscopy* **2001**, *17*, 461–476. [CrossRef]
10. Liao, C.-J.; Lin, Y.-J.; Chiang, H.; Chiang, S.-F.; Wang, Y.-H.; Jiang, C.-C. Injecting partially digested cartilage fragments into a biphasic scaffold to generate osteochondral composites in a nude mice model. *J. Biomed. Mater. Res. A* **2007**, *81*, 567–577. [CrossRef]

11. Hong, L.T.A.; Kim, Y.-M.; Park, H.H.; Hwang, D.H.; Cui, Y.; Lee, E.M.; Yahn, S.; Lee, J.K.; Song, S.-C.; Kim, B.G. An injectable hydrogel enhances tissue repair after spinal cord injury by promoting extracellular matrix remodeling. *Nat. Commun.* **2017**, *8*, 533. [CrossRef] [PubMed]
12. Chen, G.; Hoffman, A.S. Graft copolymers that exhibit temperature-induced phase transitions over a wide range of pH. *Nature* **1995**, *373*, 49–52. [CrossRef] [PubMed]
13. Lin, Z.; Gao, W.; Hu, H.; Ma, K.; He, B.; Dai, W.; Wang, X.; Wang, J.; Zhang, X.; Zhang, Q. Novel thermo-sensitive hydrogel system with paclitaxel nanocrystals: High drug-loading, sustained drug release and extended local retention guaranteeing better efficacy and lower toxicity. *J. Control. Release* **2014**, *174*, 161–170. [CrossRef] [PubMed]
14. Ruel-Gariépy, E.; Chenite, A.; Chaput, C.; Guirguis, S.; Leroux, J. Characterization of thermosensitive chitosan gels for the sustained delivery of drugs. *Int. J. Pharm.* **2000**, *203*, 89–98. [CrossRef]
15. Chenite, A.; Chaput, C.; Wang, D.; Combes, C.; Buschmann, M.D.; Hoemann, C.D.; Leroux, J.C.; Atkinson, B.L.; Binette, F.; Selmani, A. Novel injectable neutral solutions of chitosan form biodegradable gels in situ. *Biomaterials* **2000**, *21*, 2155–2161. [CrossRef]
16. Ahmadi, R.; de Bruijn, J.D. Biocompatibility and gelation of chitosan-glycerol phosphate hydrogels. *J. Biomed. Mater. Res. A* **2008**, *86*, 824–832. [CrossRef] [PubMed]
17. Roughley, P.; Hoemann, C.; DesRosiers, E.; Mwale, F.; Antoniou, J.; Alini, M. The potential of chitosan-based gels containing intervertebral disc cells for nucleus pulposus supplementation. *Biomaterials* **2006**, *27*, 388–396. [CrossRef]
18. Huang, Y.; Onyeri, S.; Siewe, M.; Moshfeghian, A.; Madihally, S.V. In vitro characterization of chitosan-gelatin scaffolds for tissue engineering. *Biomaterials* **2005**, *26*, 7616–7627. [CrossRef]
19. Cheng, Y.-H.; Yang, S.-H.; Su, W.-Y.; Chen, Y.-C.; Yang, K.-C.; Cheng, W.T.-K.; Wu, S.-C.; Lin, F.-H. Thermosensitive chitosan-gelatin-glycerol phosphate hydrogels as a cell carrier for nucleus pulposus regeneration: An in vitro study. *Tissue Eng. Part A* **2010**, *16*, 695–703. [CrossRef]
20. Gulotta, L.V.; Kovacevic, D.; Ying, L.; Ehteshami, J.R.; Montgomery, S.; Rodeo, S.A. Augmentation of tendon-to-bone healing with a magnesium-based bone adhesive. *Am. J. Sports Med.* **2008**, *36*, 1290–1297. [CrossRef]
21. Goradia, V.K.; Rochat, M.C.; Kida, M.; Grana, W.A. Natural history of a hamstring tendon autograft used for anterior cruciate ligament reconstruction in a sheep model. *Am. J. Sports Med.* **2000**, *28*, 40–46. [CrossRef] [PubMed]
22. Grana, W.A.; Egle, D.M.; Mahnken, R.; Goodhart, C.W. An analysis of autograft fixation after anterior cruciate ligament reconstruction in a rabbit model. *Am. J. Sports Med.* **1994**, *22*, 344–351. [CrossRef] [PubMed]
23. Bedi, A.; Kawamura, S.; Ying, L.; Rodeo, S.A. Differences in tendon graft healing between the intra-articular and extra-articular ends of a bone tunnel. *HSS J.* **2009**, *5*, 51–57. [CrossRef] [PubMed]
24. Smith, P.A.; Stannard, J.P.; Pfeiffer, F.M.; Kuroki, K.; Bozynski, C.C.; Cook, J.L. Suspensory Versus Interference Screw Fixation for Arthroscopic Anterior Cruciate Ligament Reconstruction in a Translational Large-Animal Model. *Arthroscopy* **2016**, *32*, 1086–1097. [CrossRef] [PubMed]
25. Yamazaki, S.; Yasuda, K.; Tomita, F.; Tohyama, H.; Minami, A. The effect of transforming growth factor-β1 on intraosseous healing of flexor tendon autograft replacement of anterior cruciate ligament in dogs. *Arthrosc. J. Arthrosc. Relat. Surg.* **2005**, *21*, 1034–1041. [CrossRef]
26. Lovric, V.; Chen, D.; Yu, Y.; Oliver, R.A.; Genin, F.; Walsh, W.R. Effects of Demineralized Bone Matrix on Tendon-Bone Healing in an Intra-articular Rodent Model. *Am. J. Sports Med.* **2012**, *40*, 2365–2374. [CrossRef]
27. Mutsuzaki, H.; Fujie, H.; Nakajima, H.; Fukagawa, M.; Nomura, S.; Sakane, M. Effect of Calcium Phosphate–Hybridized Tendon Graft in Anatomic Single-Bundle ACL Reconstruction in Goats. *Orthop. J. Sports Med.* **2016**, *4*. [CrossRef]
28. Yoon, J.P.; Chung, S.W.; Jung, J.W.; Lee, Y.S.; Kim, K.I.; Park, G.Y.; Kim, H.M.; Choi, J.H. Is a Local Administration of Parathyroid Hormone Effective to Tendon-to-Bone Healing in a Rat Rotator Cuff Repair Model? *J. Orthop. Res.* **2019**, *38*, 82–91. [CrossRef]
29. Talhouk, R.S.; Bissell, M.J.; Werb, Z. Coordinated expression of extracellular matrix-degrading proteinases and their inhibitors regulates mammary epithelial function during involution. *J. Cell Biol.* **1992**, *118*, 1271–1282. [CrossRef]
30. Bramono, D.S.; Richmond, J.C.; Weitzel, P.P.; Kaplan, D.L.; Altman, G.H. Matrix metalloproteinases and their clinical applications in orthopaedics. *Clin. Orthop. Relat. Res.* **2004**, *428*, 272–285. [CrossRef]

31. De Mos, M.; van El, B.; DeGroot, J.; Jahr, H.; van Schie, H.T.M.; van Arkel, E.R.; Tol, H.; Heijboer, R.; van Osch, G.J.V.M.; Verhaar, J.A.N. Achilles tendinosis: Changes in biochemical composition and collagen turnover rate. *Am. J. Sports Med.* **2007**, *35*, 1549–1556. [CrossRef] [PubMed]
32. Jones, G.C.; Corps, A.N.; Pennington, C.J.; Clark, I.M.; Edwards, D.R.; Bradley, M.M.; Hazleman, B.L.; Riley, G.P. Expression profiling of metalloproteinases and tissue inhibitors of metalloproteinases in normal and degenerate human achilles tendon. *Arthritis Rheum.* **2006**, *54*, 832–842. [CrossRef] [PubMed]
33. Lavagnino, M.; Arnoczky, S.P.; Egerbacher, M.; Gardner, K.L.; Burns, M.E. Isolated fibrillar damage in tendons stimulates local collagenase mRNA expression and protein synthesis. *J. Biomech.* **2006**, *39*, 2355–2362. [CrossRef]
34. September, A.V.; Cook, J.; Handley, C.J.; van der Merwe, L.; Schwellnus, M.P.; Collins, M. Variants within the COL5A1 gene are associated with Achilles tendinopathy in two populations. *Br. J. Sports Med.* **2009**, *43*, 357–365. [CrossRef]
35. Treviño, E.A.; McFaline-Figueroa, J.; Guldberg, R.E.; Platt, M.O.; Temenoff, J.S. Full-thickness rotator cuff tear in rat results in distinct temporal expression of multiple proteases in tendon, muscle, and cartilage. *J. Orthop. Res.* **2018**, *37*, 490–502. [CrossRef]
36. Bedi, A.; Kovacevic, D.; Hettrich, C.; Gulotta, L.V.; Ehteshami, J.R.; Warren, R.F.; Rodeo, S.A. The effect of matrix metalloproteinase inhibition on tendon-to-bone healing in a rotator cuff repair model. *J. Shoulder Elb. Surg.* **2010**, *19*, 384–391. [CrossRef]
37. Haslauer, C.M.; Proffen, B.L.; Johnson, V.M.; Murray, M.M. Expression of modulators of extracellular matrix structure after anterior cruciate ligament injury. *Wound Repair Regen* **2014**, *22*, 103–110. [CrossRef]
38. Demirag, B.; Sarisozen, B.; Ozer, O.; Kaplan, T.; Ozturk, C. Enhancement of tendon-bone healing of anterior cruciate ligament grafts by blockage of matrix metalloproteinases. *J. Bone Jt. Surg.* **2005**, *87*, 2401–2410.
39. Paiva, K.B.S.; Granjeiro, J.M. Bone tissue remodeling and development: Focus on matrix metalloproteinase functions. *Arch. Biochem. Biophys.* **2014**, *561*, 74–87. [CrossRef]
40. Kylmaoja, E.; Nakamura, M.; Tuukkanen, J. Osteoclasts and Remodeling Based Bone Formation. *Curr. Stem. Cell Res. Ther.* **2016**, *11*, 626–633. [CrossRef]
41. Chen, Y.-J.; Wang, C.-J.; Yang, K.D.; Kuo, Y.-R.; Huang, H.-C.; Huang, Y.-T.; Sun, Y.-C.; Wang, F.-S. Extracorporeal shock waves promote healing of collagenase-induced Achilles tendinitis and increase TGF-beta1 and IGF-I expression. *J. Orthop. Res.* **2004**, *22*, 854–861. [CrossRef] [PubMed]

© 2020 by the authors. Licensee MDPI, Basel, Switzerland. This article is an open access article distributed under the terms and conditions of the Creative Commons Attribution (CC BY) license (http://creativecommons.org/licenses/by/4.0/).

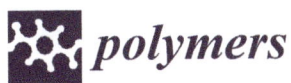

Article

Label-Free Multi-Microfluidic Immunoassays with Liquid Crystals on Polydimethylsiloxane Biosensing Chips

Yu-Jui Fan [1,2,3,†], Fu-Lun Chen [4,5,†], Jian-Chiun Liou [1], Yu-Wen Huang [1], Chun-Han Chen [1], Zi-Yin Hong [1], Jia-De Lin [6] and Yu-Cheng Hsiao [2,3,*]

1. School of Biomedical Engineering, Taipei Medical University, 250 Wuxing St., Taipei 11031, Taiwan; ray.yj.fan@tmu.edu.tw (Y.-J.F.); jcliou@tmu.edu.tw (J.-C.L.); prince99720@gmail.com (Y.-W.H.); andychen060188@gmail.com (C.-H.C.); b507105053@tmu.edu.tw (Z.-Y.H.)
2. International PhD Program for Biomedical Engineering, Taipei Medical University, 250 Wuxing St., Taipei 11031, Taiwan
3. Graduate Institute of Biomedical Optomechatronics, College of Biomedical Engineering, Taipei Medical University, 250 Wuxing St., Taipei 11031, Taiwan
4. Division of Infectious Diseases, Department of Internal Medicine, Wan Fang Hospital, Taipei Medical University, No.111, Sec. 3, Xinglong Rd., Wenshan Dist., Taipei 116, Taiwan; 96003@w.tmu.edu.tw
5. Department of Internal Medicine, School of Medicine, College of Medicine, Taipei Medical University, 250 Wuxing St., Taipei 11031, Taiwan
6. Department of Engineering Science, University of Oxford, Parks Road, Oxford OX1 3PJ, UK; geman1218@yahoo.com.tw
* Correspondence: ychsiao@tmu.edu.tw; Tel.: +886-291-937-9783
† These authors contributed equally to this work.

Received: 16 January 2020; Accepted: 30 January 2020; Published: 10 February 2020

Abstract: We developed a new format for liquid crystal (LC)-based multi-microfluidic immunoassays, hosted on a polydimethylsiloxane substrate. In this design, the orientations of the LCs were strongly affected by the interface between the four microchannel walls and surrounding LCs. When the alignment layer was coated inside a microchannel, the LCs oriented homeotropically and appeared dark under crossed polarizers. After antigens bound to the immobilized antibodies on the alignment layer were coated onto the channel walls, the light intensity of the LC molecules changed from dark to bright because of disruption of the LCs. By employing pressure-driven flow, binding of the antigen/antibody could be detected by optical signals in a sequential order. The multi-microfluidic LC biosensor was tested by detecting bovine serum albumin (BSA) and an immunocomplex of BSA antigen/antibody pairs, a protein standard commonly used in labs. We show that this multi-microfluidic immunoassay was able to detect BSA and antigen/antibody BSA pairs with a naked-eye detection limitation of −0.01 µg/mL. Based on this new immunoassay design, a simple and robust device for LC-based label-free microfluidic immunodetection was demonstrated.

Keywords: polydimethylsiloxane; microfluidic; bovine serum albumin

1. Introduction

Microfluidic immunoassays based on polydimethylsiloxane (PDMS) substrates possess several advantages such as short assay times, small volumes, and low costs [1,2]. However, a microfluidic immunoassay needs a special signal detection method and enhancement technique such as fluorescence detection or enzymatic reaction enhancement method, since the miniature size of microfluidic devices reduces the signal. Thus, antibody/antigen pairs in microfluidic devices are usually conjugated with labels like enzymes [3,4], fluorophores [5,6], nanoparticles [7,8], and so on, in order to transduce the immunobinding response into detectable signals. In addition, the biofunctionality of antibody/antigen pairs is affected upon conjugation with labels [9,10].

Recently, liquid crystals (LCs) were successfully used for label-free biodetection [11]. It was reported that the reorientation of LCs caused the immunobinding response to be more sensitive and changed the optical signals of LCs [12,13]. In addition, a protein standard, bovine serum albumin (BSA), was successfully detected using LC materials [14,15]. Such changes in the optical properties of LCs enable unique optical properties such as Bragg reflection, flexibility, and bi-stability [16–21]. In addition, the first cholesteric LC (CLC) biosensing device was proposed by Hsiao et al. in 2015 [22]. A highly sensitive color-indicating CLC biosensor was invented. However, it is unfortunate that CLC biosensors require complicated fabrication processes. In addition, LC biosensors must be confined in a well-defined space for biosensing applications such as in a cell device [23] or transmission electron microscopic (TEM) grid [24]. In addition, assembly of these LC devices requires additional complicated steps, which are time-consuming. Currently, these issues are the main obstacles to applying LC biosensors in practice. In addition, Chen et al. invented the first single-substrate CLC biosensing device [25]. It seems that a better way to resolve time-consuming detection by LC biosensors can be found.

Based on the problems mentioned above, we planned to take advantage of the fluidic property of LC materials, as well as propose an LC-based multi-microfluidic immunoassay to detect BSA antigen/antibody pairs. However, some efforts were proposed for integrating LC materials into microfluidic devices. The first LC/PDMS film to detect ethanol in microfluidic devices was recently proposed [26]. The device enables monitoring of ethanol production. However, this device is only applicable for detecting organic molecules. In addition, Liu et al. revealed an LC microfluidic-filled metal grid for detecting detergents and enzymes [27]. Thus, real-time monitoring by an LC microfluidic device has been achieved.

In this study, we developed an LC-based multi-microfluidic immunoassay chip. We show the behavior between BSA antigen/antibody pairs and LC molecules in a PDMS microchannel. We show that this device substantially differs from a typical optical cell. Since the LC molecules within a microchannel are in contact with the four alignment-coated channel walls, the orientations of the LCs are determined by both the surface functionality and the geometric dimensions of the channels. Antigens/antibodies can be detected by observing the optical appearance of LCs inside the microchannel under crossed polarizers. A highly sensitive interface between LC molecules and an alignment layer of N,N-dimethyl-n-octadecyl-3-aminopropyltrimethoxysilyl chloride (DMOAP) was employed to detect BSA concentrations. The schematic of this multi-microfluidic LC/PDMS biosensor is shown in Figure 1.

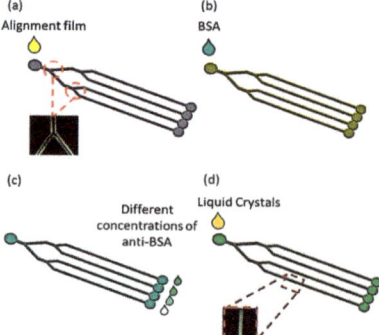

Figure 1. Schematic of multi-microfluidic liquid crystal (LC) immunoassays. The LC configuration changes from the homeotropic to the planar mode in the presence of biomolecules on DMOAP-coated channels. BSA is bovine serum albumin.

2. Materials and Methods

Single-layer and two-level cascaded microchannels were fabricated by a general PDMS-based soft lithographic fabrication process. First, a microchannel mold was fabricated on a 4-inch (10.16-cm) silicon wafer by standard photolithography using SU-8 2025 photoresist. The thickness of the mold was 25 µm. Then the PDMS base and curing agent were mixed in a 10:1 weight ratio. After allowing 30 min for degassing, the mixture was poured into the master. After baking at 65 °C for 4 h, the cured PDMS microchannel was peeled off the master. Using oxygen plasma treatment, the PDMS microchannel and a pre-cleaned glass substrate were tightly bonded together. The LC mixture used in this study was nematic E44. In addition, the multi-microfluidic channels were immersed in the DMOAP aqueous solution for 30 min to coat the aligned layer onto the inner walls of the microfluidic channel. Next, the multi-microfluidic device was rinsed with deionized water for 1 min in order to remove excess DMOAP solution. For test immobilization, BSA in the form of an aqueous solution was filled into the DMOAP-coated microfluidic channel. In addition, BSA concentrations of 1 and 0.1 mg/mL, and 10, 1, 0.1, and 0.01 µg/mL were used. Concentrations of an anti-BSA antibody of 0, 1, 10, and 100 µg/mL were also used. The empty microfluidic channel was filled with LC under volume flow rates of 5, 10, 20, and 30 µL/min through a syringe to form LC multi-microfluidic chips. A BX51 polarized optical microscope (POM; Olympus, Tokyo, Japan) equipped with a halogen light bulb as the light source was used in this study for observation. All experimental data were acquired at 26 ± 1 °C.

3. Results and Discussion

3.1. Optical Properties of Multi-Microfluidic LC Biosensors

Polarized optical images of the LC multi-microfluidic biosensors with BSA concentrations of 0 and 1 mg/mL under crossed and paralleled polarizers are shown in Figure 2. One can observe that the darkness and brightness of the microfluidic biosensor corresponded to 0 and 1 mg/mL, respectively. The vertical alignment layer of DMOAP forced the LCs to orient perpendicularly to the surface of the microfluidic channel, and the microfluidic channel without biomolecules appeared dark under crossed polarizers. However, the vertical anchoring force of the alignment layer (DMOAP) was diminished by the immobilized BSA [25], and the brightness of the microfluidic channel increased when observed under crossed polarizers, as shown in Figure 2. Moreover, the phenomenon under parallel polarizers was opposite of that under crossed polarizers. Since the contrast was better under crossed polarizers, we used crossed polarizers for the following quantification measurements. In addition, the LC material used in this report exhibited the LC phase for a wide range of temperatures. Therefore, the LC multi-microfluidic biosensor can conveniently be used in a wide range of temperatures in different settings. Polarized optical images of LC microfluidic biosensors with immobilized BSA protein of 0–1 mg/mL are also shown in Figure 3. In the absence of BSA, the LC multi-microfluidic biosensor was dark, but it became brighter with an increasing BSA concentration. The DMOAP-coated substrate caused LC molecules to be homeotropically aligned. Therefore, it was possible to measure the amount of biomolecules with multi-microfluidic LC chips. In addition, the vertical anchoring force of the alignment layer of DMOAP was diminished by the immobilized BSA layer. The LC molecules converted to a non-homeotropic state with an increasing concentration of BSA. Finally, the planar arrangement caused the biosensor to be in a bright state. In order to achieve quantitative data for LC multi-microfluidic devices, the image from POM was quantified with software [28]. Linear correlations of the intensity of microfluidic LC chips at different BSA concentrations are shown in Figure 4. A higher intensity was shown with a rising BSA concentration. The linear equation between the BSA concentration (y) and light intensity (x) was $y = 0.49x$. These experimental results proved that the LCs in the multi-microfluidic can also be used for quantitating and detecting biomolecules in a linear manner as a new label-free biosensor.

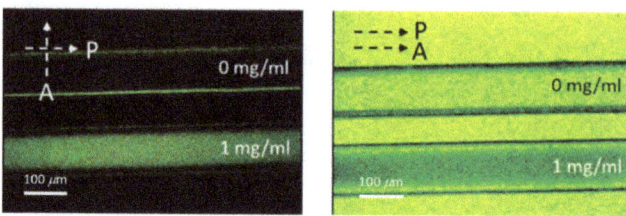

Figure 2. Optical images from a polarized optical microscope of liquid crystal (LC) multi-microfluidic biosensors at both 0 and 1 mg/mL concentrations of bovine serum albumin (BSA) under crossed and parallel polarizer conditions. P stands for polarizer and A is the analyzer.

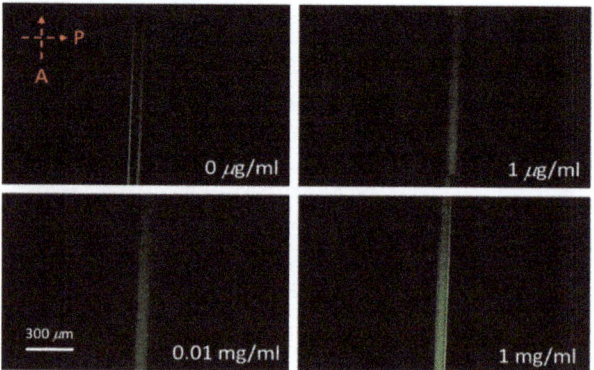

Figure 3. Polarized optical images from a polarized optical microscope of liquid crystal (LC) microfluidic biosensors with immobilized bovine serum albumin (BSA) at 0–1 mg/mL. P stands for polarizer and A is the analyzer.

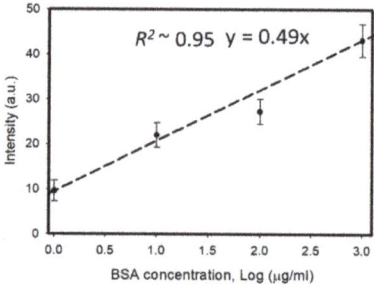

Figure 4. Linear correlations of the transmitted intensity of multi-microfluidic liquid crystal (LC) immunoassay chips at different bovine serum albumin (BSA) concentrations.

3.2. Quantitation for Immunoassay LC Microfluidic Devices

In addition, an immunoassay test of the device was also examined using both BSA and an anti-BSA antibody. Intensities of the immunoassay LC microfluidic devices immobilized with 0, 1, and 10 µg/mL concentrations of BSA and 0, 10, 100, and 1000 µg/mL concentrations of the anti-BSA antibody are shown in Figure 5. We mixed 0–1000 µg/mL of the anti-BSA antibody with identical concentrations of the BSA antigen at concentrations of 0–10 µg/mL to allow the formation of immunocomplexes between specific antigen/antibody pairs. We observed that with lower concentrations of the anti-BSA antibody (<10 µg/mL), immunocomplexes could not form between the specific antigen/antibody pairs. The intensities of the immunocomplexes at 1 and 10 µg/mL concentrations of BSA did not obviously change. When 100 and 1000 µg/mL of the anti-BSA antibody were mixed, the BSA antigen/antibody mixtures produced a much brighter state under POM. However, an excess concentration of the anti-BSA antibody strongly affected the LC arrangement, which meant that the intensity of change in the BSA immunocomplexes could not be analyzed or quantified. These results suggest that BSA immunocomplexes, compared to those with the BSA antigen or antibody alone, induced more significant disruption of the LC arrangement (Figure 5). From the experimental data, 1 µg/mL of the anti-BSA antibody was a more appropriate concentration with the BSA antigen. This method of immunodetection could thus discern between immunocomplexes and unbound antigens and antibodies. The linear correlation between the transmittance intensity of the LC-based multi-microfluidic device and different BSA concentrations of <10 µg/mL of the anti-BSA antibody is shown in Figure 6. We observed that the immunodetection limit of BSA antigen/antibody pairs was 0.01 µg/mL BSA, and that corresponded to 1 µg/mL of the anti-BSA antibody. These results proved that the linear correlation of the LC-based multi-microfluidic device can be used to detect and quantitate biomolecules or for immunodetection in a linear manner. Note that as the antigens and antibodies were complexed through multiple noncovalent interactions, such as hydrogen bonds, electrostatic interactions, van der Waals forces, and so on; 6 h of pre-drying was required to minimize these effects and improve the stability of the BSA immunocomplexes in this study. Based on the sensitivity, label-free state, multi-detection ability, and ease of manufacture, this study shows that LC multi-microfluidic chips have potential for development as a label-free, highly sensitive, cheap, multi-detection, and immunodetection biosensing technique.

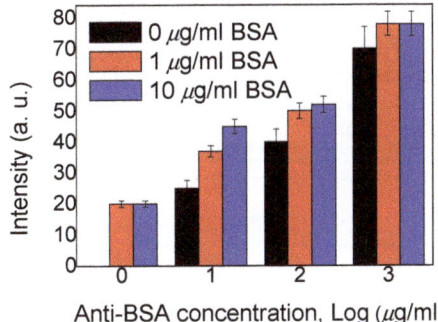

Figure 5. Intensities of immunoassay liquid crystal (LC) microfluidic chips immobilized with 0, 1, and 10 µg/mL concentrations of BSA and 0, 10, 100, and 1000 µg/mL concentrations of the anti-BSA antibody.

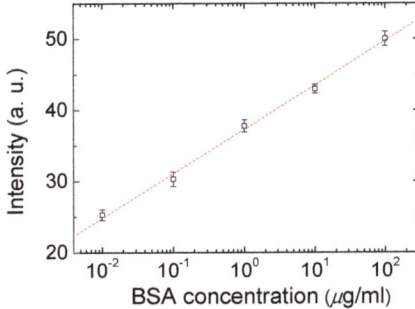

Figure 6. Transmitted intensities of multi-microfluidic liquid crystal (LC) immunoassay chips at different bovine serum albumin (BSA) concentrations mixed with a 10 µg/mL concentration of an anti-BSA antibody.

3.3. Effects of Volume Flow Rates in LC Microfluidic Devices

Volume flow rates in LC microfluidic devices are important because of the fluidity of LC molecules. Figure 7 shows the effects of different volume flow rates of LC injected into the microchannel of chips on the optical performance. Too-rapid volume flow rates (>10 µL/min) resulted in a disordered arrangement of the LCs and produced a defective optical texture [29]. When the volume flow rate was ≥30 µL/min, the too-fast volume flow rates caused a small amount of LC molecules to be left in the microchannel. Due to the disorder of the LCs under rapid volume flow rates, the intensity of the microchannels decreased when the volume flow rate increased. Thus, volume flow rates of <10 µL/min have to be used in LC-based microfluidic devices. In this study, we used a volume flow rate of 5 µL/min to perform LC injection experiments to ensure uniformity of the LCs in the microchannels.

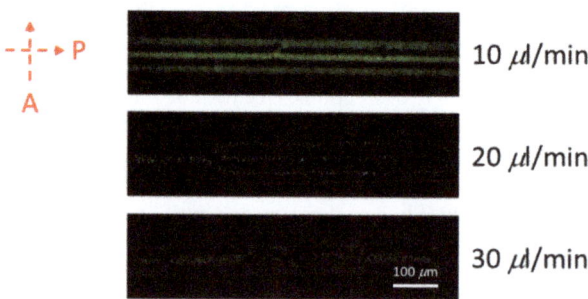

Figure 7. Optical images of a 1 mg/mL concentration of bovine serum albumin (BSA) under different volume flow rates of 10, 20, and 30 µL/min of liquid crystals (LCs) injected into the microchannel. P stands for polarizer and A is the analyzer.

3.4. Comparisons to Other Label-Free Biosensing Techniques

Moreover, many label-free biosensing techniques were proposed in the past. The most important and commonly used techniques are grating coupled interferometry (GCI) and plasmonic sensing [30–33]. The problem with both GCI and plasmonic sensors is that the measurement process is complicated. In addition, both of these sensing devices are expensive and require large equipment, which is not portable or easy to use. Compared to both label-free plasmonic and GCI techniques, our LC microfluidic device is cheaper and can be measured with a smartphone [34] or detailed spectrum. Compared to the well-known immunodetection method, our multi-microfluidic LC immunoassay chips are cheaper and easier to use. Based on the naked-eye detection property of the label-free nature, this study showed that the LC microfluidic device has potential for development as a cheap, sensitive, and portable biosensing technique for immunodetection.

4. Conclusions

A novel design of LC-based multi-microfluidic immunoassay chips was proposed. The orientation of the LCs was strongly influenced by the interface between the four channel walls and surrounding LC molecules. When a DMOAP alignment layer was coated onto the interior of the microchannel, the LCs oriented homeotropically and appeared dark under crossed polarizers. After the antigens had bound to the immobilized antibodies in the multi-microfluidic device, the appearance of the LC phase changed from dark to bright because of disruption of the LC orientation. Using pressure-driven flow, antigen/antibody binding could be detected by optical signals in a sequential order. The immunodetection limit of BSA antigen/antibody pairs was 0.01 µg/mL BSA and 1 µg/mL of the anti-BSA antibody. We show that this multi-microfluidic LC immunoassay chip is able to detect BSA and antigen/antibody BSA immunocomplexes with label-free immunodetection.

This new design of an immunoassay device provides a sensitive, cheap, multi-detection, and robust approach to LC-based immunodetection.

Author Contributions: Data curation, F.-L.C., Y.-J.F., C.-H.C., J.-C.L., Y.-W.H., and Z.-Y.H.; funding acquisition, Y.-C.H.; investigation, Y.-J.F.; project administration, Y.-C.H.; writing, Y.-C.H. and J.-D.L. All authors have read and agreed to the published version of the manuscript.

Funding: This work was financially supported by the Ministry of Science and Technology, Taiwan, under grant no. MOST108-2636-E-038-001, Taipei Medical University, Wan Fang Hospital, Taiwan under grant no. 108TMU-WFH-26, and Taipei Medical University, Taiwan, under grant no. TMU106-AE1-B49.

Conflicts of Interest: The authors declare that no conflict of interest.

References

1. Bange, A.; Halsall, H.B.; Heineman, W.R. Microfluidic immunosensor systems. *Biosens. Bioelectron.* **2005**, *20*, 2488–2503. [CrossRef] [PubMed]
2. Henares, T.G.; Mizutani, F.; Hisamoto, H. Current development in microfluidic immunosensing chip. *Anal. Chim. Acta* **2008**, *611*, 17–30. [CrossRef] [PubMed]
3. Eteshola, E.; Balberg, M. Microfluidic ELISA: On-chip fluorescence imaging. *Biomed. Microdevices* **2004**, *6*, 7–9. [CrossRef] [PubMed]
4. Yu, L.; Li, C.M.; Liu, Y.; Gao, J.; Wang, W.; Gan, Y. Flow-through functionalized PDMS microfluidic channels with dextran derivative for ELISAs. *Lab Chip* **2009**, *9*, 1243–1247. [CrossRef]
5. Bernard, A.; Michel, B.; Delamarche, E. Micromosaic immunoassays. *Anal. Chem.* **2001**, *73*, 8–12. [CrossRef]
6. Hosokawa, K.; Omata, M.; Maeda, M. Immunoassay on a power-free microchip with laminar flow-assisted dendritic amplification. *Anal. Chem.* **2007**, *79*, 6000–6004. [CrossRef]
7. Lu, Y.; Shi, W.; Qin, J.; Lin, B. Low cost, portable detection of gold nanoparticlelabeled microfluidic immunoassay with camera cell phone. *Electrophoresis* **2009**, *30*, 579–582. [CrossRef]
8. Luo, C.; Fu, Q.; Li, H.; Xu, L.; Sun, M.; Ouyang, Q.; Chen, Y.; Ji, H. PDMS microfluidic device for optical detection of protein immunoassay using gold nanoparticles. *Lab Chip* **2005**, *5*, 726–729. [CrossRef]
9. Thorek, D.L.J.; Elias, D.R.; Tsourkas, A. Comparative analysis of nanoparticleantibody conjugations: Carbodiimide versus click chemistry. *Mol. Imaging* **2009**, *8*, 221–229. [CrossRef]
10. Shrestha, D.; Bagosi, A.; Szöllosi, J.; Jenei, A. Comparative study of the three different fluorophore antibody conjugation strategies. *Anal. Bioanal. Chem.* **2012**, *404*, 1449–1463. [CrossRef]
11. Gupta, V.K.; Skaife, J.J.; Dubrovsky, T.B.; Abbott, N.L. Optical amplification of ligand-receptor binding using liquid crystals. *Science* **1998**, *279*, 2077–2080. [CrossRef] [PubMed]
12. Aliño, V.J.; Yang, K.L. Using liquid crystals as a readout system in urinary albumin assays. *Analyst* **2011**, *136*, 3307–3313. [CrossRef]
13. Chen, C.H.; Yang, K.L. Liquid crystal-based immunoassays for detecting hepatitis B antibody. *Anal. Biochem.* **2012**, *421*, 321–323. [CrossRef] [PubMed]
14. Kim, S.R.; Abbott, N.L. Rubbed films of functionalized bovine serum albumin as substrates for the imaging of protein–receptor interactions using liquid crystals. *Adv. Mater.* **2001**, *13*, 1445–1449. [CrossRef]
15. Clare, B.H.; Abbott, N.L. Orientations of nematic liquid crystals on surfaces presenting controlled densities of peptides: Amplification of protein–peptide binding events. *Langmuir* **2005**, *21*, 6451–6461. [CrossRef] [PubMed]
16. Hsiao, Y.C.; Tang, C.Y.; Lee, W. Fast-switching bistable cholesteric intensity modulator. *Opt. Express* **2011**, *19*, 9744–29749. [CrossRef] [PubMed]
17. Hsiao, Y.C.; Wu, C.Y.; Chen, C.H.; Zyryanov, V.Y.; Lee, W. Electro-optical device based on photonic structure with a dual-frequency cholesteric liquid crystal. *Opt. Lett.* **2011**, *36*, 2632–2634. [CrossRef]
18. Hsiao, Y.C.; Hou, C.T.; Zyryanov, V.Y.; Lee, W. Multichannel photonic devices based on tristable polymer-stabilized cholesteric textures. *Opt. Express* **2011**, *19*, 23952–23957. [CrossRef]
19. Hsiao, Y.C.; Zou, Y.H.; Timofeev, I.V.; Zyryanov, V.Y.; Lee, W. Spectral modulation of a bistable liquid-crystal photonic structure by the polarization effect. *Opt. Mater. Express* **2013**, *3*, 821–828. [CrossRef]
20. Hsiao, Y.C.; Wang, H.T.; Lee, W. Thermodielectric generation of defect modes in a photonic liquid crystal. *Opt. Express* **2014**, *22*, 3593–3599. [CrossRef]

21. Hsiao, Y.C.; Lee, W. Electrically induced red, green, and blue scattering in chiral-nematic thin films. *Opt. Lett.* **2015**, *40*, 1201–1203. [CrossRef] [PubMed]
22. Hsiao, Y.C.; Sung, Y.C.; Lee, M.J.; Lee, W. Highly sensitive color-indicating and quantitative biosensor based on cholesteric liquid crystal. *Biomed. Opt. Express* **2015**, *6*, 5033–5038. [CrossRef] [PubMed]
23. Lowe, A.M.; Ozer, B.H.; Bai, Y.P.; Bertics, J.; Abbott, N.L. Design of surfaces for liquid crystal-based bioanalytical assays. *ACS Appl. Mater. Interfaces* **2010**, *2*, 722–731. [CrossRef] [PubMed]
24. Bi, X.; Yang, K.L. Liquid crystals decorated with linear oligopeptide FLAG for applications in immunobiosensors. *Biosens. Bioelectron.* **2010**, *26*, 107–111. [CrossRef] [PubMed]
25. Chen, F.-L.; Fan, Y.-J.; Lin, J.-D.; Hsiao, Y.-C. Label-free, color-indicating, and sensitive biosensors of cholesteric liquid crystals on a single vertically aligned substrate. *Biomed. Opt. Express* **2019**, *10*, 4636–4642. [CrossRef]
26. Sutarlie, L.; Yang, K.L. Monitoring spatial distribution of ethanol in microfluidic channels by using a thin layer of cholesteric liquid crystal. *Lab Chip* **2011**, *11*, 4093–4098. [CrossRef]
27. Liu, Y.; Cheng, D.; Lin, I.H.; Abbott, N.L.; Jiang, H. Microfluidic sensing devices employing in situ-formed liquid crystal thin film for detection of biochemical interactions. *Lab Chip* **2012**, *12*, 3746–3753. [CrossRef]
28. Hsiao, Y.C. Liquid crystal-based tunable photonic crystals for pulse compression and signal enhancement in multiphoton fluorescence. *Opt. Mater. Express* **2016**, *6*, 1929–1934. [CrossRef]
29. Hsiao, Y.C.; Lee, W. Polymer stabilization of electrohydrodynamic instability in non-iridescent cholesteric thin films. *Opt. Express* **2015**, *23*, 22636–22642. [CrossRef]
30. Kozma, P.; Hamori, A.; Cottier, K.; Kurunczi, S.; Horvath, R. Grating coupled interferometry for optical sensing. *Appl. Phys. B* **2009**, *97*, 5–8. [CrossRef]
31. Kozma, P.; Hamori, A.; Kurunczi, S.; Cottier, K.; Horvath, R. Grating coupled optical waveguide interferometer for label-free biosensing. *Sens. Actuators B Chem.* **2011**, *155*, 446–450. [CrossRef]
32. Abdulhalim, I. Optimized guided mode resonant structure as thermooptic sensor and liquid crystal tunable filter. *Chin. Opt. Lett.* **2009**, *7*, 667. [CrossRef]
33. Abdulhalim, I. Plasmonic Sensing using Metallic Nano-Sculptured Thin Films. *Small* **2014**, *10*, 3499. [CrossRef] [PubMed]
34. Chen, F.-L.; Fan, Y.-J.; Lin, J.-D.; Hsiao, Y.-C. Label-free biosensor of reflective polymer-stabilized nanostructure thin-films. **2020**. Submitted.

© 2020 by the authors. Licensee MDPI, Basel, Switzerland. This article is an open access article distributed under the terms and conditions of the Creative Commons Attribution (CC BY) license (http://creativecommons.org/licenses/by/4.0/).

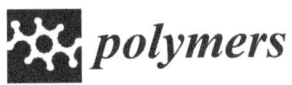

Article

Porcine Collagen–Bone Composite Induced Osteoblast Differentiation and Bone Regeneration In Vitro and In Vivo

Eisner Salamanca [1,†], Chia Chen Hsu [1,2,†], Wan Ling Yao [1,*], Cheuk Sing Choy [3,4,†], Yu Hwa Pan [1,5,6,7], Nai-Chia Teng [1,8] and Wei-Jen Chang [1,2,*]

1. School of Dentistry, College of Oral Medicine, Taipei Medical University, Taipei 110, Taiwan; D204103004@tmu.edu.tw (E.S.); m204103002@tmu.edu.tw (C.C.H.); shalom.dc@msa.hinet.net (Y.H.P.); dianaten@tmu.edu.tw (N.-C.T.)
2. Dental Department, Taipei Medical University, Shuang-Ho hospital, Taipei 235, Taiwan
3. Department of Community Medicine, En Chu Kong Hospital, New Taipei City 237, Taiwan; prof.choy@gmail.com
4. Yuanpei University of Medical technology, Hsin Chu, Taipei 300, Taiwan
5. Department of Dentistry, Chang Gung Memorial Hospital, Taipei 106, Taiwan
6. Graduate Institute of Dental & Craniofacial Science, Chang Gung University, Taoyuan 333, Taiwan
7. School of Dentistry, College of Medicine, China Medical University, Taichung 404, Taiwan
8. Dental Department, Taipei Medical University Hospital, Taipei 110, Taiwan
* Correspondence: yaoyao061637@gmail.com (W.L.Y.); cweijen1@tmu.edu.tw (W.-J.C.); Tel.: +886-2-2736-1661 (ext. 5148) (W.-J.C.); Fax: +886-2-2736-2295 (W.-J.C.)
† Authors did the equal contribution to this work.

Received: 22 November 2019; Accepted: 5 December 2019; Published: 4 January 2020

Abstract: Due to autogenous bone limitations, some substitute bone grafts were developed. Collagenated porcine graft (CPG) is able to regenerate new bone, although the number of studies is insufficient, highlighting the need for future studies to better understand the biomaterial. In order to understand better CPG's possible dental guided bone regeneration indications, the aim of this work was to determine CPG's biological capacity to induce osteoblast differentiation in vitro and guided bone regeneration in vivo, whilst being compared with commercial hydroxyapatite and beta tricalcium phosphate (HA/β-TCP) and porcine graft alone. Cell cytotoxicity (WST-1), alkaline phosphatase activity (ALP), and real-time polymerase chain reaction (qPCR) were assessed in vitro. Critical size defects of New Zealand white rabbits were used for the in vivo part, with critical size defect closures and histological analyses. WST-1 and ALP indicated that CPG directly stimulated a greater proliferation and confluency of cells with osteoblastic differentiation in vitro. Gene sequencing indicated stable bone formation markers, decreased resorption makers, and bone remodeling coupling factors, making the transition from osteoclast to osteoblast expression at the end of seven days. CPG resulted in the highest new bone regeneration by osteoconduction in critical size defects of rabbit calvaria at eight weeks. Nonetheless, all biomaterials achieved nearly complete calvaria defect closure. CPG was found to be osteoconductive, like porcine graft and HA/β-TCP, but with higher new bone formation in critical size defects of rabbit calvaria at eight weeks. CPG can be used for different dental guided bone regeneration procedures; however, further studies are necessary.

Keywords: porcine graft; collagen type I; biological apatite; chemical properties

1. Introduction

In the fields of periodontics, endodontics, implantology, maxillofacial surgery, and orthopedics, performing bone regeneration procedures on a daily basis is a common clinical practice. For these kinds of treatments, the patient's bone, also known as autogenous bone, is the gold standard and, hence, the first choice. From a macro perspective, autogenous bone is insufficient to cover all bone-grafting procedures due to its limited availability. Over two million bone-grafting procedures are estimated to be performed annually worldwide [1]. In addition, the use of autogenous bone is associated with an 8%–39% risk of major and minor complications [2] related to the donor site harvesting procedure, such as infection, wound drainage, hematomas, reoperation, prolonged pain, sensory loss, and keloids. At a minimum, the patient must undergo a second skin incision, in addition to extended anesthesia time and hospital stay [3]. For these reasons, it is important to have various options available to augment, expand, or substitute autologous bone grafts [4]. Technological evolution and an improved understanding of bone biology led to the development of several bone graft substitutes that are currently available for the treatment of large cancellous voids [5]. Depending on their origin, these substitutes can be classified as allografts, xenografts, and synthetic grafts. All of them provide an osteoconductive or structural framework for bone ingrowth that can be used for bone regeneration while reconstructing significant bone voids [6].

Bone substitutes of natural origin exhibit greater osteoconductivity compared to derivative substitutes [5]. Xenografts are good alternatives because of the unlimited supply of available material and because they eliminate the need for extra procedures to harvest bone. Bovine xenograft was popular in the 1960s but fell into disfavor because of the potential for iatrogenic transmission of prion-related diseases to patients treated with this product [7]. However, the risk declined due to the adoption of appropriate heat treatments to deproteinate bone particles [8,9]. Currently, porcine xenograft is more commonly used mainly because pigs have a genotype close to that of humans [10]. Porcine bone has a similar chemical composition and structure as human bone [4,11,12], and it was reported to be biocompatible and highly osteoconductive in clinical studies. Based on the histomorphometric and osteoblast gene expression profiles, the grafting of corticocancellous porcine bone to heal extraction sockets in humans confirmed these characteristics [13].

Collagen type I is part of the chemical composition and structure of human bone. Previous studies demonstrated that collagen has some superior properties to other materials, including hemostatic functions that allow early wound stabilization, hemostatic properties to attract fibroblasts, and semi-permeability, which facilitates nutrient transfer. The major drawback of collagen is its fast biodegradation due to the enzymatic activity of macrophages and polymorphonuclear leucocytes [14,15]. Furthermore, collagen exhibits elasticity and mechanical toughness, while calcium phosphate exhibits mechanical stiffness to withstand compressive loads [16]. Therefore, it is important to incorporate collagen within bone particle grafts to mimic biological human bone for better bone regeneration outcomes. Multiple studies using collagenated porcine graft (CPG) for different augmentation surgeries showed that new bone formation and implant placement can be safely carried out later [17]. Furthermore, CPG was proven to reduce bone loss when compared to naturally healing sockets [18,19]. Although the available data show good results, the number of studies is insufficient, highlighting the need for future studies to better understand CPG. This work was conducted to determine the biological capacity of CPG for the induction of osteoblast differentiation in vitro and guided bone regeneration in vivo in comparison with commercial hydroxyapatite and beta tricalcium phosphate HA/β-TCP and porcine graft alone. The aim was to gain a better understanding of the material and its possible indications in dental guided bone regeneration procedures.

2. Materials and Methods

2.1. Graft Materials

CPG is a homogeneous plug consisting of purified porcine type I collagen mixed with porcine bone graft at a weight ratio of 30:70 (Sunmax Biotechnology, Tainan, Taiwan. Figure 1). The preparation process for the material was previously reported in another study performed by the same laboratory [7,20]. Briefly, pig bones were settled in water to remove soft tissue with two different ramp rates until the temperature reached 800 and 1000 °C. Afterward, bone proteins were etched and removed with 0.1–0.5 M hydrochloric acid (HCl) for 10 min. Next, the bones were rinsed, heat-dried, filtered to a particle size of 500–1000 μm, and sterilized using γ-rays. Later, porcine type I collagen was mixed with porcine bone graft, poured into a mold, and allowed to freeze-dry until CPG was formed. HA/β-TCP used was a biphasic ceramic material (MBCP™; Biomatlante, Vigneux de Bretagne, France) consisting of 60% HA and 40% β-TCP with complete interconnected porosity of 70%. The material comprised macropores of >10 μm and micropores of <10 μm in a 2:1 ratio, and it was sintered at temperatures >700 °C [21].

Figure 1. Porcine type I collagen mixed with porcine bone graft at a weight ratio of 30:70. Particles can be seen evenly distributed through the collagen.

2.2. Cell Culture and Seeding

MG-63 osteoblast-like cells were purchased from Bioresource Collection and Research Center (BCRC, Hsinchu, Taiwan). The cells were expanded in Dulbecco's modified Eagle's medium (DMEM; HyClone, Logan, UT, USA) supplemented with L-glutamine (4 mmol/L), 10% fetal bovine serum, and 1% penicillin–streptomycin (HyClone, Logan, UT, USA) at 37 °C in a humidified atmosphere containing 95% air and 5% CO_2. The confluent cells were sub-cultured to the next passage using 0.05% trypsin– ethylenediaminetetraacetic acid (EDTA) for up to the fourth passage. Once 90% confluence cell density was reached, the concentration was adjusted to 1×10^4 cells/mL, and the cells were aliquoted into 24-well Petri dishes (Nunclon; Nunc, Roskilde, Denmark). On the same day, the DMEM medium was mixed with CPG, porcine graft, or HA/β-TCP in a 1 g/10 mL concentration. Twenty-four hours later, the medium in the test well was removed and substituted for test media consisting of the previously described DMEM + CPG, porcine graft, or HA/β-TCP. The DMEM medium first described was used for the control wells. The medium was changed every three days for all wells.

2.3. Cell Cytotoxicity

Cell cytotoxicity was assessed one, three, and seven days after DMEM + particle grafts were added to the test wells. The test was performed according to the Cell Proliferation Reagent manufacturer's instructions (WST-1 Kit, Roche Applied Science, Mannheim, Germany). The assay principle is based on the conversion of the tetrazolium salt WST-1 into a colored dye by mitochondrial dehydrogenase enzymes. In brief, the medium from the cells, prepared as previously described, was replaced with 500 μL of fresh medium. Later, the cells were moved to a 96-well microtiter plate (5×10^4 cells/well) in a final volume of 100 μL of culture medium in the absence of any remaining particle grafts. Afterward, the cells were incubated for 24 h, and 10 μL of WST-1 reagent was added to each well, followed by incubation for 4 h in the same standard culture conditions. Next, the plate was placed on a shaker for 1 min to mix the contents. Subsequently, the absorbance of the samples was measured using a Multiskan™ GO Microplate Spectrophotometer (Thermo Fisher Scientific, Waltham, MA, USA) at the optical density (OD) of 420–480 nm with a reference wavelength of 650 nm. The percentage cytotoxicity was calculated from the following equation: % cytotoxicity = $(100 \times (control - sample))/control$ [22].

2.4. Assessment of Cell Morphology by Fluorescence Microscopy

Image analysis-based cell morphology in two-dimensional (2D) culture was evaluated by staining with phalloidin and with the DNA-binding dye DAPI (4′,6-diamidine-2′-phenylindole dihydrochloride) according to manufacturer's protocols (D9542 Sigma-Aldrich Deisenhofen, Germany). Staining was performed one, three, and seven days after cell culture. Cells were fixed using 4% paraformaldehyde. Substrates were then washed with buffer to remove loosely bound cells prior to imaging. Morphological results of the cell culture were visualized at one, three, and seven days using an inverted fluorescence microscope (Leica, DFC 7000T, Wetzlar, Germany) with excitation at 488 nm and detection at 530 nm fluorescin diacetate (FDA, green) and 620 nm propidium iodide (PI, red). Visual field was selected randomly [23].

2.5. Alkaline Phosphatase Activity

After cell culture and seeding, alkaline phosphatase activity was measured on the first, third, and seventh days. Cells were washed twice with phosphate-buffered saline (PBS). PBS was removed using suction, and 300 µL of Triton X-100 (BioShop, Canada Inc. Burlington, Ontario, Cannada) was added at a concentration of 0.05%. To induce rupture, the cells were subjected to three cycles of 5 min at 37 °C and 5 min at −4 °C, and the samples were later placed into 96-well plates. The alkaline phosphatase (ALP) activity was determined by following the Thermo Scientific 1-Step p-nitrophenyl phosphate disodium salt (PNPP) manufacturer's instructions. PNPP was supplied pre-mixed with substrate buffer and ready to use at room temperature. The 1-Step PNPP was gently mixed. Next, 100 µL of the mixture was added to each well in the 96-well plate and mixed thoroughly by gently agitating the plate, and the 96-well plates were incubated at room temperature for 30 min. To stop the reaction, 50 µL of 3 M NaOH was added and mixed thoroughly by gently agitating the plate. The absorbance of each well was measured at 405 nm using a Multiskan™ GO Microplate Spectrophotometer (Thermo Fisher Scientific, Waltham, MA, USA). The enzymatic activity was normalized to the total protein concentration using bovine serum albumin (Roche, Basel, Switzerland). The measurement of protein was done by using the standard Bradford method (Sigma). The ALP activity was expressed as µm/mg protein/assay time. The comparison was done by plotting the OD intensity [24].

2.6. Real-Time Polymerase Chain Reaction (qPCR)

For RNA processing, cell culture and seeding were performed after zero, three, and seven days. Total RNA was extracted using the Novel Total RNA Mini Kit (NovelGene, Molecular Biotech, Taiwan) according to the manufacturer's instructions. For RNA processing, the cells were trypsinized, harvested, and resuspended in 100 µL of PBS, before being subjected to cell lysis by adding 400 µL of natural rubber and 4 µL of S-mercaptoethanol to the sample. RNA binding was later performed with 400 µL of 70% ethanol and centrifuged at 13,000 rpm. Afterward, the sample was washed and eluted with 50 µL of RNase-free water [25,26].

Expression was quantified using qPCR. Gene expression levels were normalized to the expression of the housekeeping gene glyceraldehyde 3-phosphate dehydrogenase (*GAPDH*). Analysis results were expressed as time-course gene changes relative to the cell's genes cultured in DMEM only, and the calibrator sample representing the amount of transcript was expressed on day zero [27]. Real-time PCR was performed using 2 µL of complementary DNA (cDNA) in a 20-µL reaction volume with the LightCycler®96 Instrument and application software (Roche Molecular Systems, Inc., Pleasanton, CA, USA), and Fast SYBR™ Green Master Mix (Thermo Fisher Scientific, Vilnius, Lithuania). The temperature profile of the reaction was 95 °C for 10 min, followed by 40 cycles of denaturation at 95 °C for 15 s, annealing at 60 °C for 60 s, and extension at 72 °C for 30 s. Quantification was performed using the delta–delta calculation method. Forward and reverse primer sequences were designed using Primer-BLAST from the United States (US) National Library of Medicine and are listed in Table 1 [28].

Table 1. Primer sequences for real-time polymerase chain reaction.

Gene Symbol	Forward primer sequence (5′ > 3′)	Reverse primer sequence (5′ > 3′)
ALP	AGCCTTCCTGAAAGAGGATTGG	GCCAGTACTTGGGGTCTTTCT
OC	TCCTTTGGGGTTTGGCCTAC	CCAGCCTCCAGCACTGTTTA
RANKL	ACTGGCCTCTCACCTTTTCTG	AGCCATCCACCATCGCTTTC
CR	TTGCTGCCCGCAATTTATGA	TGCTGGCAAGATACTCAGGT
OPG	CTGGAACCCCAGAGCGAAAT	GCCTCCTCACACAGGGTAAC
RANK	GAAGGTGGACTGGCTACCAC	TTTCCTTCCCCTCCCCAGAA
GAPDH	CCTCCTGTTCGACAGTCAGC	CCTAGCCTCCCGGGTTTCTC

2.7. In Vivo Test

All animal experiments were approved by the Taipei Medical University animal ethics committee and performed following the laboratory animal center guidelines using a protocol previously described by the same laboratory [7,20]. Twenty adult male New Zealand white rabbits (mean age: 12 weeks, mean weight 3.2 kg) were housed in cages at 19 °C and 55% humidity and fed standard rabbit chow and water ad libitum. Anesthesia was administered using an intramuscular injection of Zoletil 50 (50 mg/ml. Vibac Laboratories, Carros, France) at 15 mg/kg into the gluteal region, and surgery was performed on the animals after 10 min of sedation. The calvaria region was then shaved, draped, and sterilized using iodine. Local anesthesia with 1.8 mL of 2% lidocaine with epinephrine at 1/100,000 was injected as a hemostatic in the region. Subsequently, a 2-cm longitudinal midline vertical skin and periosteum incision was made. Calvaria bones were exposed, and four (6 × 3 mm) critical calvaria defects were created bilaterally in the parietal and frontal bones using a low-speed trephine bur with continuous saline cooling (3I Implant Innovation, Palm Beach Gardens, FL, USA). Each defect was filled with a different material: porcine graft, CPG, or HA/β-TCP. Only the control defect was left unfilled. The rabbits were kept in cages under surveillance for the first 24 h and then examined every three days for two weeks and weekly thereafter. The animals (n = 5 per group) were sacrificed at two, four, six, and eight weeks after surgery. Euthanasia was performed by CO_2 asphyxiation 10 min after intramuscular injection of Zoletil 50 (50 mg/mL) at 15 mg/kg into the gluteal region. Subsequently, defects were recovered and prepared for micro-computed tomography (micro-CT) and histological analysis.

2.8. Micro-CT Scanning Cortical Defect Closure

Sample blocks were prepared in formalin and micro-computed tomography (micro-CT) scanning analyses was performed within two weeks using Skyscan 1076 (Skyscan, Antwerp, Belgium). After setting the micro-CT images, coronal images of the upper and lower peripheral areas of the defect were saved in the database, and three-dimensional (3D) morphological analyses were performed for cortical defect closure measurements. Thus, binary selections of samples from the morphometric analyses were made according to grayscale density between units of 20 and 80. The analyses were performed using Skyscan 1076 data-viewer software according to the manufacturer's instructions.

2.9. Histological Analysis

Masson's trichrome (MT) staining was performed for the detection of new bone formation and the remaining graft material. The harvested samples were decalcified in 10% EDTA for two weeks, embedded in paraffin, and cut from a sagittal perspective using a microtome (Leica RM2145 microtome, Wetzlar, Germany) to obtain the two most central sections (5 μm thick) for each defect [29]. Histomorphometric studies were performed by an experienced investigator blinded to the experiment. For histomorphometric analysis, images of three regions of interest (ROIs), as well as the borders and middle of the defect (2 mm × 2 mm), were captured using a Leica/Aperio ScanScope System. To measure new bone formation and the remaining graft material, ImageJ software (National Institutes of Health; Bethesda, MD, USA) was used [7].

2.10. Statistical Analysis

Comparisons were made between the CPG, porcine graft, HA/β-TCP, and control groups at different time periods. A non-parametric one-sample Wilcoxon test was used to identify differences between groups, while a nonparametric Kruskal–Wallis test followed by a Mann–Whitney test was used to identify statistical differences between the different time points. For all resulting parameters, the mean ± standard deviation was used, and statistical significance was set at $p < 0.05$ for all the tests. All data analyses were performed with Microsoft Excel Professional Plus 2016 (Microsoft Software, Redmond, WA, USA).

3. Results

3.1. Cell Culture and Organization

The cell cytotoxicity assay results are presented in Figure 2, showing that there was a clear trend of greater proliferation in the DMEM + CPG medium than in the other media at one, three, and seven days. Additionally, at all time points, DMEM combined with CPG or HA/β-TCP was non-toxic and more viable than DMEM + porcine graft and DMEM alone. Cells cultivated in DMEM + porcine graft presented similar toxicity and viability as the control medium without any statistically significant differences in any of the time points. Statistically significant differences ($p < 0.05$) are outlined in Figure 2. The morphology of the incubated MG-63 osteoblast-like cells in all media was similar over seven days when observed with light microscopy. There were no apparent specific changes in the first 24 h, but changes in cell number and confluency were observed after 72 h, with well-attached cells presenting an elongated shape (Figure 3).

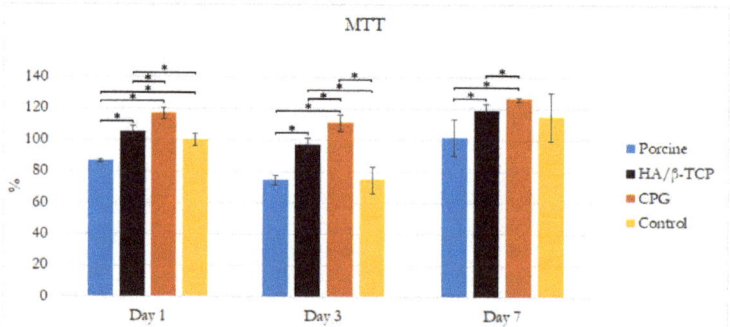

Figure 2. Effects of collagenated porcine graft (CPG), porcine graft, and hydroxyapatite and beta tricalcium phosphate (HA/β-TCP) on MG-63 osteoblast-like cell viability. Cells were incubated from days 0–7. Results are expressed as percentages of control and are the means ± standard error. Statistically significant differences were set at $p < 0.05$ and are indicated by asterisks (*).

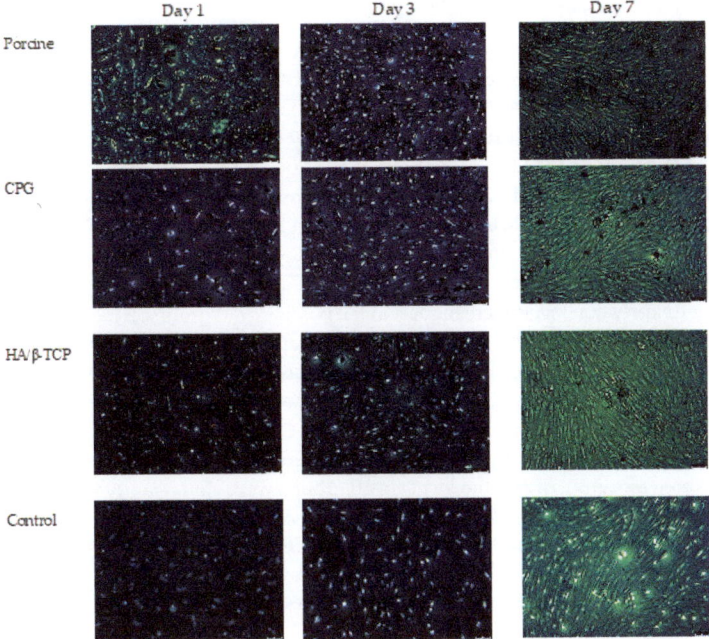

Figure 3. Fluorescence microscopy of MG-63 cells co-cultured with particle graft medium. Changes in MG-63 cell morphology and spreading at one, three, and seven days. (magnification 20×).

3.2. Alkaline Phosphatase Assay

The effect of CPG, porcine graft, and HA/β-TCP on cellular differentiation was assessed by measuring ALP activity. As shown in Figure 4, there was a higher increase in ALP expression for cells cultured in DMEM + CPG medium over the other cells on days three and seven ($p < 0.05$). These results strongly suggest that CPG directly stimulates the osteoblastic differentiation of MG-63 osteoblast-like cells in vitro. In addition, DMEM mixed with porcine graft or HA/β-TCP increased the cells' ALP activity in a time-dependent manner, better than DMEM alone after seven days. Statistically significant differences ($p < 0.05$) are outlined in Figure 4.

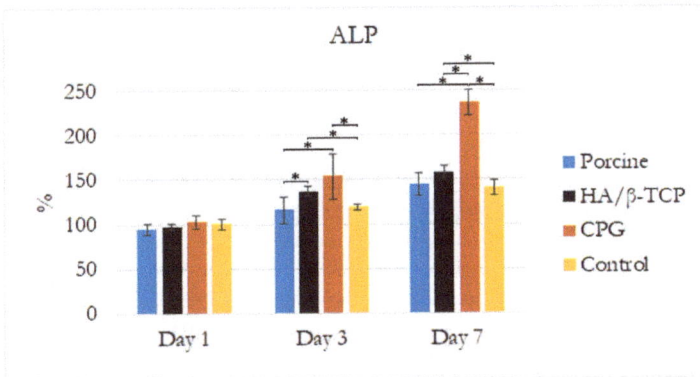

Figure 4. Effects of CPG, porcine graft, and HA/β-TCP on alkaline phosphatase (ALP) activity of MG-63 osteoblast-like cells after seven days. Results are expressed as percentages of control and are the means ± standard error. Statistically significant differences were set at $p < 0.05$ and are indicated by asterisks (*).

3.3. Real-Time Polymerase Chain Reaction (qPCR)

To characterize the phenotypes of the differentiated MG-63 and confirm their osteoblast phenotype, the gene expression of cells cultured in DMEM control or combined with CPG, porcine graft, or HA/β-TCP was analyzed by performing qPCR on days three and seven of co-cultivation. The results were compared to those for cells cultured in DMEM only, which was used as the calibrator sample on day zero. The targeted gene markers for bone formation were *ALP* and osteocalcin (*OC*). *CR* and receptor activator of nuclear factor Kappa B (*RANK*) were assigned as markers for bone resorption, and osteoprotegerin (*OPG*), RANK ligand (*RANKL*), and *RANK* were selected as markers for bone remodeling coupling. The expression of *GAPDH* was determined to confirm the usage of similar amounts of RNA for RT-PCR.

The *ALP* gene expression levels were highest in the porcine graft group on day three and were nearly unaltered by day seven. The *ALP* levels in the CPG and HA/β-TCP groups were consistently better than that in the control group, with an increase in the expression over seven days. The expression levels of *OC* were increased in all biomaterial media on day three compared to that of the control, with the highest increase observed in the porcine graft group. On day seven, *OC* exhibited a reduction in the porcine graft and CPG groups, with no major change in the HA/β-TCP group. The expression level of *CR* was about the same for all three test groups, and the levels in all groups were significantly higher than the control level on day three. This changed on day seven, when significantly lower expression of *CR* was observed compared to that of the control, as well as a significantly lower expression with CPG and HA/β-TCP. The porcine graft group showed reduced *CR* expression on day seven but still had higher expression than the control group, with no significant differences between the two groups (Figure 5).

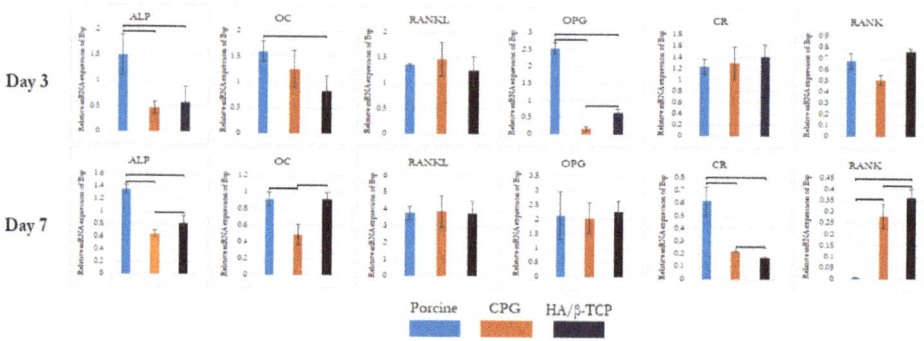

Figure 5. MG-63 gene expression on days three and seven. CPG, porcine graft, and HA/β-TCP in MG-63 osteoblast-like cells induced greater osteoblast gene expression than the control. Statistically significant differences were observed between the groups each day and between days compared to the control, which was established as day zero. Statistically significant differences were set at $p < 0.05$ and are indicated by asterisks (*).

When the groups were compared at each time point, a similar trend was observed for the *RANKL* expression. On day three, all groups had higher *RANKL* expression than the control, with nearly the same levels ($p < 0.05$). On day seven, a sharp increase in *RANKL* level was demonstrated by all groups with the same behavior, with a statistically significant difference compared to the control on day three. In contrast, *RANK* showed a gradual decrease over time. On day three, all groups had similar *RANK* levels with significant difference from the control, followed by a marked reduction on day seven, with all groups having less *RANK* expression than the control group ($p < 0.05$). *OPG* was expressed in the porcine graft group at a higher level with a sharp increase over the rest of the groups on day three. During the same period, the CPG and HA/β-TCP groups had significantly lower and nearly the same expression levels, respectively, compared to the control ($p < 0.05$). A significant difference was observed on day seven, with the CPG and HA/β-TCP groups reaching the same level as the porcine graft group, which showed a similar pattern as on day three. All groups had statistically significant differences compared to the control on day 7 ($p < 0.05$) (Figure 5).

3.4. Defect Closure

During all study periods, defect closure measured in percentage was found to increase, resulting in the closing of most defects. The control defects did not close entirely and had the least amount of closure among the groups, having only 72.19% ± 24.08% and 64.63% ± 19.04% upper and lower closure overall over eight weeks (Figure 6, Table 2). Due to the incomplete closure of the control defects, they were considered critical size defects. Active treatment of CPG resulted in the best defect closure, with 99.42% ± 1.3% upper and 99.49% ± 1.14% lower closure of the defects ($p < 0.05$). When observed at eight weeks, the HA/β-TCP group (97.19% ± 1.65% upper and 97.73% ± 1.57% lower) was better than the porcine graft group (94.66% ± 5.3% upper and 95.64% ± 3.91% lower), with nearly all the defects closed. Differences between the use of graft materials vs. natural healing were statistically significant (Table 2) in favor of the active biomaterials.

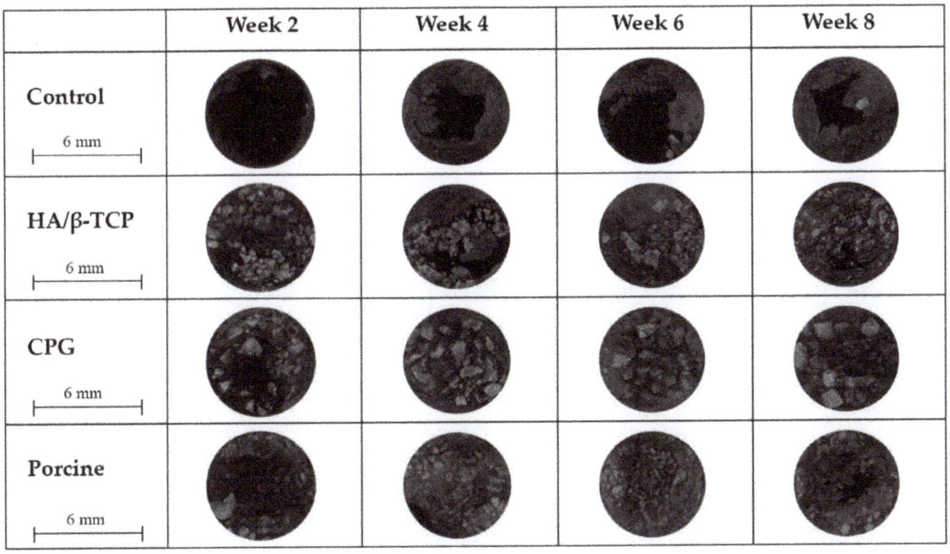

Figure 6. Upper view of critical size defects with non-closure in control defects vs. nearly complete closure with CPG, porcine graft, and HA/β-TCP over eight weeks.

Table 2. Percentage of calvaria defect closure. CPG—collagenated porcine graft.

	Defect Upper Side				Defect Lower Side			
	2 Weeks	4 Weeks	6 Weeks	8 Weeks	2 Weeks	4 Weeks	6 Weeks	8 Weeks
Control	6.25 ± 7.99 *	57.95 ± 17.04 *	56.98 ± 35.34 *	72.19 ± 24.08 *	14.67 ± 11.48 *	55.22 ± 15.8 *	49.28 ± 31.01 *	64.63 ± 19.04 *
HA/β-TCP	65.56 ± 20.66	84.22 ± 6.95	93.48 ± 13.31	97.19 ± 1.65	61.74 ± 7.99	78.73 ± 12.45	93.99 ± 12.18	97.73 ± 1.57
CPG	73.65 ± 12.77	99.17 ± 1.06	100 ± 0	99.42 ± 1.3 ¶	52.88 ± 17.95	98.84 ± 1.28	100 ± 0	99.49 ± 1.14 ¶
Porcine	63.32 ± 18.44	94.58 ± 10.31	95.59 ± 7.37	94.66 ± 5.3	65.02 ± 20.42	95.56 ± 8.57	95.98 ± 6.77	95.64 ± 3.91

Results are expressed as percentages of control and are the means ± standard error. * indicates $p < 0.05$ when all groups were better than control within the same timeframe, whilst ¶ shows better defect closure for CPG compared to the other groups in the same timeframe ($p < 0.05$).

3.5. Histological and Histomorphometric Analyses

Histological results revealed that the defect sites exhibited variable degrees of healing between the defects filled with graft materials. All defect sites had a significant difference from the control defects except for the first two weeks when histologic examination revealed the greatest concentrations of immature bone in the borders of all defects. The control defects had the lowest bone formation of 7.07% ± 5.51% followed by HA/β-TCP defects with 10.30% ± 5.65% of new bone and 21.4% ± 7.7% of graft filling the defects. Porcine graft defects with 20.24% ± 3% had the highest amount of new bone formation with only 18.6% ± 3% of the graft filling the defect. The CPG defects had 16.58% ± 3.27% of new bone formation, but double the graft amount filled the defects (37.3% ± 4.7%). In all defects, the discrete presence of some inflammatory multinucleated giant cells and blood capillaries with red blood cells were observed. In addition to this moderate inflammatory response, some fibroblasts were observed around the graft particles in the CPG, porcine graft, and HA/β-TCP defects, indicating a more intense inflammatory response than in the control defects (Figures 7–9).

A reduced number of inflammatory cells along with new bone formation and connective tissue were found at four weeks in all defects. Grafted defects presented biomaterial particles surrounded by some demineralized bone particles. In the porcine defects, 32.89% ± 4.37% of new bone was easily distinguishable from the 13.6% ± 2.4% of grafted particles. Similarly, 22.11% ± 3.43% new bone and 22.55% ± 4.76% osteoid formation were observed in the control and CPG defects, with 39.6% ± 3.4% incomplete resorbed graft particles in the CPG defects. HA/β-TCP defects had only 15.4% ± 2.2% of non-reabsorbed particles grafts, but the particles were well integrated and in complete continuity with the 24.72% ± 5.56% of new bone tissue formation (Figures 7–9).

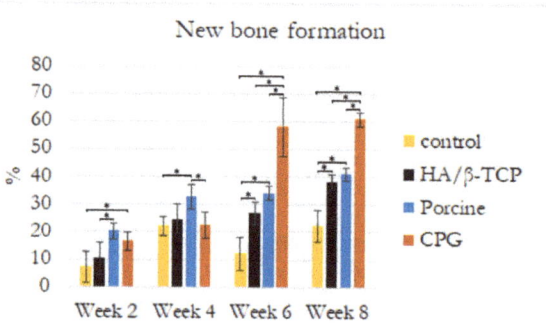

Figure 7. Histological new bone formation percentage at two, four, six, and eight weeks. Bone area was measured in the regions of interest (2 mm × 2 mm) at the borders and center of the most central part of the rabbit calvaria defects. Asterisks (*) indicate statistically significant differences ($p < 0.05$).

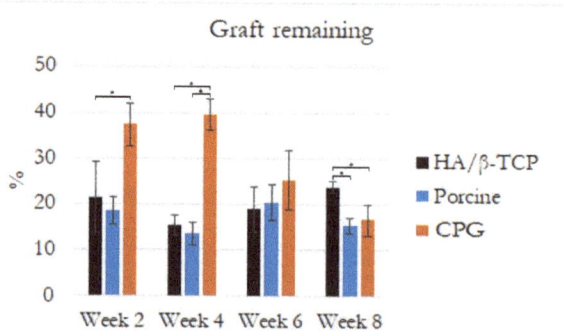

Figure 8. Histological graft remaining percentage at two, four, six, and eight weeks. Graft particle area was measured in the regions of interest (2 mm × 2 mm) at the borders and center of the most central part of the rabbit calvaria defects. Asterisks (*) indicate statistically significant differences with more resorption in CPG at the end of eight weeks ($p < 0.05$).

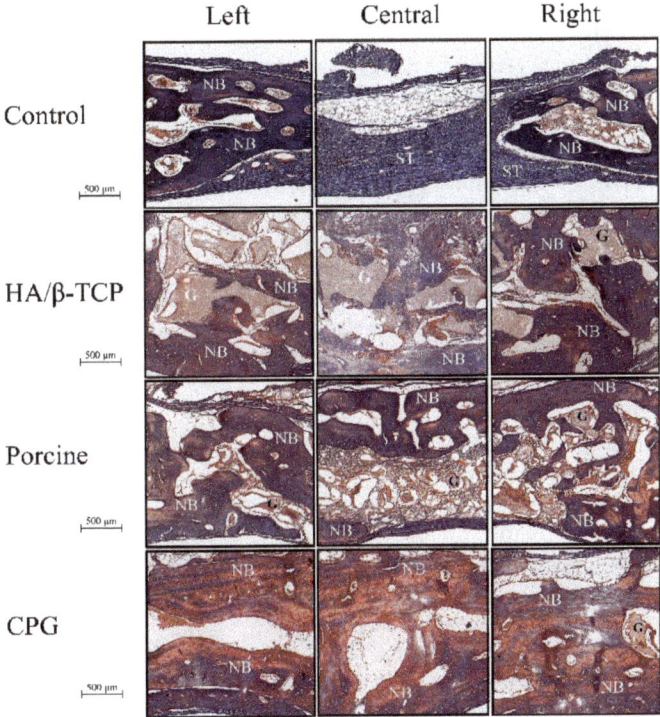

Figure 9. Histological micrographs of mid-sagittal section's regions of interest at eight weeks. ST: soft tissue, G: graft particles, NB: new bone, OB: old bone.

In all samples at six weeks, trabecular bone was observed over the entire grafted area, with the highest amount of new bone (57.93% ± 10.51%) on the CPG defects grafted to 25.4% ± 6.5% of the remaining material particles. In the porcine defects, 20.4% ± 3.8% of particles were surrounded by 34.25% ± 2.34% of regenerated thin new bone trabeculae, while others were partially in direct contact with the woven bone. The same characteristics were found within the 26.90% ± 4.13% new bone formation in HA/β-TCP defects, which sometimes presented newly formed bone areas inside the 18.9% ± 4.9% remnant particles (Figures 7–9).

At eight weeks, no evidence of foreign-body reaction was observed in any of the samples, and there was no evidence of necrotizing reaction. In many areas of grafted defects, wide osteocyte lacunae and rims of osteoblasts were visible with active deposition of osteoid matrix. This deposition formed 60.92% ± 2.46% of new bone and 16.6% ± 3.3% of particles in CPG defects. In the porcine graft group, the defects consisted of 41.16% ± 2.17% newly formed bone, while the HA/β-TCP defects contained 38.21% ± 2.67% newly formed bone. Furthermore, these new bones were in close and tight contact with 15.4% ± 1.7% and 23.8% ± 1.3% of the particle remnants in the porcine and HA/β-TCP defects, respectively. Similar to the grafted defects, the tissue corresponded in structure and morphology to new bone tissue in the control defects with 22.10% ± 5.81% new bone formation (Figures 7–9).

4. Discussion

The emphasis of this study was to determine CPG's biological capacity to induce osteoblast differentiation in vitro and guided bone regeneration in vivo. Gene sequencing indicated stable bone formation markers, decreased resorption makers, and bone remodeling coupling factors, making the transition from osteoclast to osteoblast expression at the end of seven days. CPG resulted in the highest guided bone regeneration in critical size defects of rabbit calvaria at eight weeks, and this was achieved by osteoconduction. CPG was slightly better than or similar to commercial HA/β-TCP and porcine graft, with better in vitro results and new bone formation. Nonetheless, all biomaterials achieved nearly complete calvaria defect closure. CPG was found to be safe, effective, and osteoconductive. Within the characteristics presented in this study, CPG can be indicated for augmentation or reconstructive treatment of alveolar ridge defects, filling in periodontal defects, defects after root resection, apicocectomy, and cystectomy, extraction sockets to enhance preservation of the alveolar ridge, maxillary sinus floor lifting, and periodontal defects in conjunction with products intended for guided tissue regeneration and guided bone regeneration.

In the cytotoxicity test and fluorescence microscopy, there was a clear trend of greater proliferation and confluency of cells, with good attachment and elongated characteristics in the DMEM + CPG medium than in the other media during the seven days of the test (Figures 2 and 3). These results demonstrated that CPG graft material, due to the presence of collagen, is less toxic and more viable than the other materials and control (Figures 2 and 3). The WST-1 cell proliferation reagent used in the present cytotoxicity assay is similar to the reagent used in the 3-(4,5-dimethylthiazol-2-yl)-2,5-diphenyltetrazolium bromide (MTT) assay in principle, as both measure the metabolic activity of viable cells. However, the WST-1 assay produces water-soluble formazan, eliminating the additional solubilization step. Thus, the WST-1 assay should be considered a more rapid alternative to the MTT assay [22]. ALP activity is essential for bone mineralization and is considered a useful biochemical marker for bone formation [30]. During the same seven-day period when the cytotoxicity test was conducted, MG-63 osteoblast-like cells cultured in DMEM + CPG had higher ALP activity, strongly suggesting that osteoblastic-like cells differentiate by depositing a mineralized extracellular matrix characteristic of bone tissue in vitro ($p < 0.05$) [31] (Figure 3).

The present study demonstrated that porcine graft increases osteogenic differentiation in the osteoblast-like MG-63 cell line more than the control, evidenced by higher expression levels of bone formation markers including *ALP* and *OC* on days three and seven. Osteogenic differentiation proceeds through different developmental stages characterized by specific markers including early markers such as *ALP* and late markers such as *OC*. Additionally, porcine graft induced a decrease in the expression of the *CR* gene, a bone resorption marker and regulator of calcium metabolism via specific receptors [32]. *RANKL/RANK/OPG* expression levels on days three and seven indicated that the cells cultured in DMEM + porcine graft had more osteoblast-related gene expression than osteoclasts. CPG and HA/β-TCP, similar to the porcine graft, induced increased osteogenic differentiation in osteoblast-like cells compared to the control. *ALP* and *OC* gene expression was lower in the CPG and HA/β-TCP groups than in the porcine graft group on day three. *CR* gene expression in the CPG and HA/β-TCP groups after seven days approached the level observed in the porcine graft group. The levels of *RANKL/RANK/OPG* in the CPG and HA/β-TCP groups indicated a slightly higher osteoclast-related gene expression than the levels in the porcine graft group. The *RANKL/RANK/OPG* pathway is important for understanding the signaling between osteoblasts and osteoclasts. Osteoblast cells express *RANKL*, which binds to its receptor, *RANK*, on the surface of osteoclasts and their precursors. This regulates the differentiation of precursors into multinucleated osteoclasts and the activation of osteoclasts. OPG is secreted by osteoblasts and osteogenic stromal stem cells and protects the skeleton from excessive bone resorption by binding to *RANKL* and preventing it from interacting with *RANK* [33].

MG63 cells were chosen because they are well characterized as immature osteoblasts, and they were applied as a tool for studying differentiation processes [31,34]. The results for *ALP* gene expression did not correspond to the biochemical analysis results for ALP activity due to the lower cell proliferation found in the cytotoxicity test. This indicated that a higher number of cells cultured with CPG in the ALP biochemical analysis were more able to deposit a mineralized extracellular matrix than cells cultured in porcine and HA/β-TCP media. However, cells cultured in porcine medium had higher osteoblast gene markers, indicating that the porcine graft induced cells into a more mature osteoblast differentiation than CPG and HA/β-TCP. Despite these differences, all three graft materials were superior to the control.

The rabbit calvaria defect animal model was chosen because the rabbit calvarial bone has similar characteristics to those of the jawbone: the intramembranous embryological origin, the presence of two cortical layers separated by cancellous bone, and the physiology of bone repair [35]. All rabbit calvaria defects were considered critical size defects because they failed to heal during the animal's lifetime in the study, making it an adequate animal model for testing CPG osteoconductivity, biocompatibility, and degradation in comparison to porcine graft and HA/β-TCP. Animal models for bone-grafting studies should be cheap, easy to obtain and handle, and adequate for the creation of large bone cavities [35]. The histomorphometric results in the present study demonstrated defect sites with variable degrees of healing between the defects filled with graft materials, and all of them were significantly different compared to the control defects. At two weeks, the CPG, porcine, and HA/β-TCP defect groups exhibited a moderate inflammatory response, some fibroblasts around the graft particles, and more intense inflammatory responses compared to the control defects. This is in agreement with Figueiredo et al.'s revelation that a xenogeneic graft mixed with collagen type I and an alloplastic material conformed to hydroxyapatite and two secondary phases of α- and β-tricalcium phosphate, similar to the ones used in our study, did not cause severe inflammation in the in vivo inflammatory response test [36]. At eight weeks, wide osteocyte lacunae and active deposition of osteoid matrix on the rims of osteoblasts were observed in many areas of the grafted defects. This deposition was caused by 60.92% ± 2.46% of new bone in the CPG defects and was in close contact with only 16.6% ± 3.3% of the remaining particles. The values were higher than those observed for the porcine and HA/β-TCP defects with 41.16% ± 2.17% and 38.21% ± 2.67% of newly formed bone, respectively. All grafted materials had higher new bone tissue than the control defects, which had only 22.10% ± 5.81% new bone formation (Figures 7–9). Similar results were found by Nannmark et al. in a study on rabbit maxillary defects evaluating bone tissue response to prehydrated and collagenated porcine bone graft with or without collagen gel. The group found that both materials showed bone formation directly on the particles with typical osteoblastic seams. Over an eight-week period, the bone area increased and the prehydrated and collagenated porcine bone graft, both mixed with and without collagen gel, was resorbed by osteoclasts. Remodeling was also observed with the formation of osteons within the particles. They concluded that prehydrated and collagenated porcine bone graft exhibited osteoconductive properties and was resorbed with time [37]. Rabbit animal models are sometimes considered the first step before larger animal studies. For particulate porcine bone mix and porcine corticocancellous collagenate prehydrated bone mix for bone regeneration, larger animal studies were performed by surgically creating bone defects around implants in sheep. The findings of this study, in agreement with ours, also revealed new bone formation around the graft particles [38].

Because CPG mimics human bone architecture, it was used in different types of human studies, including successful buccal bone augmentation around immediate implants with and without flap elevation [39]. It was also reported to be non-problematic and predictable in terms of clinical success for a case series of implant placement in fresh extraction sockets and simultaneous osteotome sinus floor elevation [40]. Furthermore, CPG was compared to porcine bone mix collagen with autologous bone and a 50:50 mixture for bone formation in sinus augmentation procedures using histological evaluation after two months. The researchers concluded that collagenated porcine bone alone or in combination with autologous bone is biocompatible and osteoconductive and can be successfully used in sinus augmentation procedures [10].

5. Conclusions

CPG had the capacity to induce MG-63 cells into mature osteoblast differentiation in vitro with higher viability and ALP activity than porcine and HA/β-TCP media. The gene sequence in cells cultivated with CPG and HA/β-TCP indicated stable bone formation markers, decreased resorption markers, and bone remodeling coupling factors, making the transition to mature osteoblast expression at the end of seven days. CPG was found to be osteoconductive, similar to porcine graft and HA/β-TCP, but with higher new bone formation in critical size defects of rabbit calvaria after eight weeks. CPG can be used for different dental bone regeneration procedures; however, further studies are necessary.

Author Contributions: Conceptualization, E.S., C.C.H., and W.-J.C.; methodology, W.L.Y., E.S., and C.C.H.; software, C.S.C.; validation, N.-C.T. and Y.-H.P.; formal analysis, C.S.C.; investigation, Y.H.P.; resources, W.-J.C.; data curation, C.C.H. and E.S.; writing—original draft preparation, C.S.C.; writing—review and editing, W.-J.C. and E.S.; visualization, N.C.T.; supervision, N.-C.T.; project administration, Y.-H.P.; funding acquisition, C.S.C. and W.-J.C. All authors have read and agreed to the published version of the manuscript.

Funding: This study was supported by the En Chu Kong Hospital grant ECKHW10713 and the Council of Agriculture grant 107AS-1.2.5-ST-a5.

Conflicts of Interest: The authors declare no conflict of interest.

References

1. Campana, V.; Milano, G.; Pagano, E.; Barba, M.; Cicione, C.; Salonna, G.; Lattanzi, W.; Logroscino, G. Bone substitutes in orthopaedic surgery: From basic science to clinical practice. *J. Mater. Sci. Mater. Med.* **2014**, *25*, 2445–2461. [CrossRef] [PubMed]
2. Van der Stok, J.; Van Lieshout, E.M.; El-Massoudi, Y.; Van Kralingen, G.H.; Patka, P. Bone substitutes in the Netherlands—A systematic literature review. *Acta Biomater.* **2011**, *7*, 739–750. [CrossRef] [PubMed]
3. Wang, W.; Yeung, K.W. Bone grafts and biomaterials substitutes for bone defect repair: A review. *Bioact. Mater.* **2017**, *2*, 224–247. [CrossRef] [PubMed]
4. Finkemeier, C.G. Bone-grafting and bone-graft substitutes. *JBJS* **2002**, *84*, 454–464. [CrossRef] [PubMed]
5. Faour, O.; Dimitriou, R.; Cousins, C.A.; Giannoudis, P.V. The use of bone graft substitutes in large cancellous voids: Any specific needs? *Injury* **2011**, *42*, S87–S90. [CrossRef]
6. Blank, A.T.; Riesgo, A.M.; Gitelis, S.; Rapp, T.B. Bone grafts, substitutes, and augments in benign orthopaedic conditions: Current concepts. *Bull. NYU Hosp. Jt. Dis.* **2017**, *75*, 119.
7. Salamanca, E.; Hsu, C.C.; Huang, H.M.; Teng, N.C.; Lin, C.T.; Pan, Y.H.; Chang, W.J. Bone regeneration using a porcine bone substitute collagen composite in vitro and in vivo. *Sci. Rep.* **2018**, *8*, 984. [CrossRef]
8. Murugan, R.; Rao, K.P.; Kumar, T.S. Heat-deproteinated xenogeneic bone from slaughterhouse waste: Physico-chemical properties. *Bull. Mater. Sci.* **2003**, *26*, 523–528. [CrossRef]
9. Etok, S.E.; Valsami-Jones, E.; Wess, T.J.; Hiller, J.C.; Maxwell, C.A.; Rogers, K.D.; Manning, D.A.; White, M.L.; Lopez-Capel, E.; Collins, M.J.; et al. Structural and chemical changes of thermally treated bone apatite. *J. Mater. Sci.* **2007**, *42*, 9807–9816. [CrossRef]
10. Cassetta, M.; Perrotti, V.; Calasso, S.; Piattelli, A.; Sinjari, B.; Iezzi, G. Bone formation in sinus augmentation procedures using autologous bone, porcine bone, and a 50: 50 mixture: A human clinical and histological evaluation at 2 months. *Clin. Oral Implant. Res.* **2015**, *26*, 1180–1184. [CrossRef]
11. Sandor, G.; Lindholm, T.; Clokie, C. Bone regeneration of the cranio-maxillofacial and dento-alveolar skeletons in the framework of tissue engineering. *Top. Tissue Eng.* **2003**, *7*, 1–46.
12. Figueiredo, M.; Henriques, J.; Martins, G.; Guerra, F.; Judas, F.; Figueiredo, H. Physicochemical characterization of biomaterials commonly used in dentistry as bone substitutes—comparison with human bone. *J. Biomed. Mater. Res. Part B Appl. Biomater. Off. J. Soc. Biomater. Jpn. Soc. Biomater. Aust. Soc. Biomater. Korean Soc. Biomater.* **2010**, *92*, 409–419. [CrossRef] [PubMed]
13. Crespi, R.; Capparé, P.; Romanos, G.E.; Mariani, E.; Benasciutti, E.; Gherlone, E. Corticocancellous porcine bone in the healing of human extraction sockets: Combining histomorphometry with osteoblast gene expression profiles in vivo. *Int. J. Oral Maxillofac. Implant.* **2011**, *26*, 866–872.
14. Schwarz, F.; Rothamel, D.; Herten, M.; Sager, M.; Becker, J. Angiogenesis pattern of native and cross-linked collagen membranes: An immunohistochemical study in the rat. *Clin. Oral Implant. Res.* **2006**, *17*, 403–409. [CrossRef] [PubMed]
15. Salamanca, E.; Tsai, C.Y.; Pan, Y.H.; Lin, Y.T.; Huang, H.M.; Teng, N.C.; Lin, C.T.; Feng, S.W.; Chang, W.J. In vitro and in vivo study of a novel porcine collagen membrane for guided bone regeneration. *Materials* **2016**, *9*, 949. [CrossRef]

16. Patel, P.P.; Buckley, C.; Taylor, B.L.; Sahyoun, C.C.; Patel, S.D.; Mont, A.J.; Mai, L.; Patel, S. and Freeman, J.W. Mechanical and Biological Evaluation of a Hydroxyapatite-Reinforced Scaffold for Bone Regeneration. *J. Biomed. Mater. Res. Part A* **2019**, *107*, 732–741. [CrossRef]
17. Pagliani, L.; Andersson, P.; Lanza, M.; Nappo, A.; Verrocchi, D.; Volpe, S.; Sennerby, L. A collagenated porcine bone substitute for augmentation at Neoss implant sites: A prospective 1-year multicenter case series study with histology. *Clin. Implant Dent. Relat. Res.* **2012**, *14*, 746–758. [CrossRef]
18. Barone, A.; Toti, P.; Quaranta, A.; Alfonsi, F.; Cucchi, A.; Negri, B.; Di Felice, R.; Marchionni, S.; Calvo-Guirado, J.L.; Covani, U.; et al. Clinical and Histological changes after ridge preservation with two xenografts: Preliminary results from a multicentre randomized controlled clinical trial. *J. Clin. Periodontol.* **2017**, *44*, 204–214. [CrossRef]
19. Barone, A.; Toti, P.; Menchini-Fabris, G.B.; Derchi, G.; Marconcini, S.; Covani, U. Extra oral digital scanning and imaging superimposition for volume analysis of bone remodeling after tooth extraction with and without 2 types of particulate porcine mineral insertion: A randomized controlled trial. *Clin. Implant Dent. Relat. Res.* **2017**, *19*, 750–759. [CrossRef]
20. Salamanca, E.; Lee, W.F.; Lin, C.Y.; Huang, H.M.; Lin, C.T.; Feng, S.W.; Chang, W.J. A novel porcine graft for regeneration of bone defects. *Materials* **2015**, *8*, 2523–2536. [CrossRef]
21. Le GUEHENNEC, L.; Goyenvalle, E.; Aguado, E.; Pilet, P.; D'Arc, M.B.; Bilban, M.; Spaethe, R.; Daculsi, G. MBCP®biphasic calcium phosphate granules and tissucol®fibrin sealant in rabbit femoral defects: The effect of fibrin on bone ingrowth. *J. Mater. Sci. Mater. Med.* **2005**, *16*, 29–35. [CrossRef] [PubMed]
22. Ngamwongsatit, P.; Banada, P.P.; Panbangred, W.; Bhunia, A.K. WST-1-based cell cytotoxicity assay as a substitute for MTT-based assay for rapid detection of toxigenic Bacillus species using CHO cell line. *J. Microbiol. Methods* **2008**, *73*, 211–215. [CrossRef] [PubMed]
23. Yoo, C.-K.; Jeon, J.-Y.; Kim, Y.-J.; Kim, S.-G.; Hwang, K.-G. Cell attachment and proliferation of osteoblast-like MG63 cells on silk fibroin membrane for guided bone regeneration. *Maxillofac. Plast. Reconstr. Surg.* **2016**, *38*, 17. [CrossRef] [PubMed]
24. Sila-Asna, M.; Bunyaratvej, A.; Maeda, S.; Kitaguchi, H.; Bunyaratavej, N. Osteoblast differentiation and bone formation gene expression in strontium-inducing bone marrow mesenchymal stem cell. *Kobe J. Med. Sci.* **2007**, *53*, 25–35.
25. Bimboim, H.; Doly, J. A rapid alkaline extraction procedure for screening recombinant plasmid DNA. *Nucleic Acids Res.* **1979**, *7*, 1513–1523. [CrossRef]
26. Vogelstein, B.; Gillespie, D. Preparative and analytical purification of DNA from agarose. *Proc. Natl. Acad. Sci. USA* **1979**, *76*, 615–619. [CrossRef]
27. Livak, K.J.; Schmittgen, T.D. Analysis of relative gene expression data using real-time quantitative PCR and the 2-$\Delta\Delta$CT method. *Methods* **2001**, *25*, 402–408. [CrossRef]
28. Sollazzo, V.; Palmieri, A.; Scapoli, L.; Martinelli, M.; Girardi, A.; Alviano, F.; Pellati, A.; Perrotti, V.; Carinci, F. Bio-Oss®acts on Stem cells derived from Peripheral Blood. *Oman Med. J.* **2010**, *25*, 26. [CrossRef]
29. Kim, J.Y.; Ahn, G.; Kim, C.; Lee, J.S.; Lee, I.G.; An, S.H.; Yun, W.S.; Kim, S.Y.; Shim, J.H. Synergistic Effects of Beta Tri-Calcium Phosphate and Porcine-Derived Decellularized Bone Extracellular Matrix in 3D-Printed Polycaprolactone Scaffold on Bone Regeneration. *Macromol. Biosci.* **2018**, *18*, 1800025. [CrossRef]
30. Vincent, D.H.; Trivedi, M.K.; Branton, A.; Trivedi, D.; Nayak, G.; Mondal, S.C.; Jana, S. Influenced of Biofield Energy Healing Treatment on Vitamin D3 for the Assessment of Bone Health Parameters in MG-63 cells. *viXra* **2018**. viXra:1807.0213.
31. Ongaro, A.; Pellati, A.; Bagheri, L.; Rizzo, P.; Caliceti, C.; Massari, L.; De Mattei, M. Characterization of notch signaling during osteogenic differentiation in human osteosarcoma cell line MG63. *J. Cell. Physiol.* **2016**, *231*, 2652–2663. [CrossRef]
32. Cappagli, V.; Potes, C.S.; Ferreira, L.B.; Tavares, C.; Eloy, C.; Elisei, R.; Sobrinho-Simões, M.; Wookey, P.J.; Soares, P. Calcitonin receptor expression in medullary thyroid carcinoma. *PeerJ* **2017**, *5*, 3778. [CrossRef] [PubMed]
33. Boyce, B.F.; Xing, L. The Rankl/Rank/Opg Pathway. *Curr. Osteoporos. Rep.* **2007**, *5*, 98–104. [CrossRef] [PubMed]
34. Ashley, J.W.; Ahn, J.; Hankenson, K.D. Notch signaling promotes osteoclast maturation and resorptive activity. *J. Cell. Biochem.* **2015**, *116*, 2598–2609. [CrossRef] [PubMed]

35. Cavalcanti, S.C.S.X.B.; Pereira, C.L.; Mazzonetto, R.; de Moraes, M.; Moreira, R.W.F. Histological and histomorphometric analyses of calcium phosphate cement in rabbit calvaria. *J. Cranio-Maxillofac. Surg.* **2008**, *36*, 354–359. [CrossRef]
36. Figueiredo, A.; Coimbra, P.; Cabrita, A.; Guerra, F.; Figueiredo, M. Comparison of a xenogeneic and an alloplastic material used in dental implants in terms of physico-chemical characteristics and in vivo inflammatory response. *Mater. Sci. Eng. C* **2013**, *33*, 3506–3513. [CrossRef]
37. Nannmark, U.; Sennerby, L. The bone tissue responses to prehydrated and collagenated cortico-cancellous porcine bone grafts: A study in rabbit maxillary defects. *Clin. Implant Dent. Relat. Res.* **2008**, *10*, 264–270. [CrossRef]
38. Scarano, A.; Lorusso, F.; Ravera, L.; Mortellaro, C.; Piattelli, A. Bone regeneration in iliac crestal defects: An experimental study on sheep. *BioMed Res. Int.* **2016**, *2016*, 4086870. [CrossRef]
39. Covani, U.; Cornelini, R.; Barone, A. Buccal bone augmentation around immediate implants with and without flap elevation: A modified approach. *Int. J. Oral Maxillofac. Implant.* **2008**, *23*, 25.
40. Barone, A.; Cornelini, R.; Ciaglia, R.; Covani, U. Implant placement in fresh extraction sockets and simultaneous osteotome sinus floor elevation: A case series. *Int. J. Periodontics Restor. Dent.* **2008**, *28*, 3.

© 2020 by the authors. Licensee MDPI, Basel, Switzerland. This article is an open access article distributed under the terms and conditions of the Creative Commons Attribution (CC BY) license (http://creativecommons.org/licenses/by/4.0/).

Article

Fabrication of PLLA/C$_3$S Composite Membrane for the Prevention of Bone Cement Leakage

Tsai-Hsueh Leu [1,2], Yang Wei [3], Yi-Shi Hwua [4], Xiao-Juan Huang [3], Jung-Tang Huang [1,*] and Ren-Jei Chung [3,*]

1. Department of Mechanical Engineering, College of Mechanical & Electrical Engineering, National Taipei University of Technology (Taipei Tech), Taipei 10608, Taiwan; DAZ90@tpech.gov.tw
2. Department of Orthopedics, Taipei City Hospital, Renai Branch, Taipei 10629, Taiwan
3. Department of Chemical Engineering and Biotechnology, National Taipei University of Technology (Taipei Tech), Taipei 10608, Taiwan; wei38@mail.ntut.edu.tw (Y.W.); xj72baby@gmail.com (X.-J.H.)
4. Department of Medical Imaging and Radiological Sciences, Central Taiwan University of Science and Technology, Taichung 40601, Taiwan; yshwua@ctust.edu.tw
* Correspondence: jthuang@ntut.edu.tw (J.-T.H.); rjchung@ntut.edu.tw (R.-J.C.); Tel.: +(886-2)-2771-2171 (ext. 2547) (R.-J.C.)

Received: 29 October 2019; Accepted: 28 November 2019; Published: 30 November 2019

Abstract: Kyphoplasty is an important treatment for stabilizing spine fractures due to osteoporosis. However, leakage of polymethyl-methacrylate (PMMA) bone cement during this procedure into the spinal canal has been reported to cause many adverse effects. In this study, we prepared an implantable membrane to serve as a barrier that avoids PMMA cement leakage during kyphoplasty procedures through a hybrid composite made of poly-l-lactic acid (PLLA) and tricalcium silicate (C$_3$S), with the addition of C$_3$S into PLLA matrix, showing enhanced mechanical and anti-degradation properties while keeping good cytocompatibility when compared to PLLA alone and most importantly, when this material design was applied under standardized PMMA cement injection conditions, no posterior wall leakage was observed after the kyphoplasty procedure in pig lumbar vertebral bone models. Testing results assess its effectiveness for clinical practice.

Keywords: kyphoplasty; polymethyl-methacrylate bone cement; barrier; poly-l-lactic acid; tricalcium silicate

---

1. Introduction

Osteoporosis is a bone disease with the symptoms of chronic pain in the waist or back, or neck pain due to a compression fracture, significantly affecting the life quality of the elderly [1,2]. To improve the cure rate and reduce low back pain and other symptoms of osteoporosis, percutaneous kyphoplasty (PKP) has been developed as a new treatment in which a gap was prepared in a fractured vertebral body using a balloon, followed by an injection of bone cement to restore the vertebral height [3,4].

Although kyphoplasty is an effective surgical treatment for an immediate reduction of pain and deformity that can accompany vertebral compression fractures, in certain cases, however, defects or clefts in the vertebral body may cause the cement leakage into the epidural space, paraspinal soft tissues, or disc space [5,6]. These cement materials, typically consisting of polymethyl methacrylate (PMMA) which offers mechanical stability, becomes a problem when the leakages of unreacted PMMA monomer liquid before cement polymerization in the bone bed occurred, which may cause remote organ embolism or local chemical or compress symptoms and even the thermal injury [7–9].

To avoid this problem, an eggshell technique was developed from Greene et al. to prevent the cement leakage with a small amount of cement applied as a protective eggshell injected firstly into the created cavity after balloon inflation during the kyphoplasty procedure [10]. Following which the balloon is re-inserted and inflated, creating a cement shell or a cement membrane around the inner walls of the created cavity, and another batch of cement was mixed and injected into the remaining cavity thereafter with limited cement leakage possibility from initial cement setting. Although this double cement application with PMMA as the cement anti-leakage shell or membrane has demonstrated its efficacy in clinical trial during the kyphoplasty procedure [11], the long term complications might be devastating due to PMMA presence on the bone cement surface [9], which makes the development of a new material design for anti-leakage membrane necessary. For example, Tetsushi Taguchi et al. had fabricated the reactive poly(vinyl alcohol) (PVA) membrane for the prevention of bone cement leakage with good potential for implantable balloon kyphoplasty, but further investigations on its clinical efficiency are still in progress [12].

On the other hand, polymeric membranes fabricated from a single material more often have limited biological performance compared to the use of hybrid biomaterials composed of biodegradable synthetic polymers and inorganic materials, with the hybrid biomaterials fitting better to bone tissue engineering applications due to their similar compositions to natural bones [13,14]. For example, biodegradable polymers such as poly(L-lactic acid) (PLLA) (L-lactic acid isomer of polylactic acid (PLA), an aliphatic thermoplastic polyester obtained by polymerizing lactide monomers) have been used for scaffold design doped with dicalcium silicate (C_2S) nanoparticles as an ideal candidate for novel bone graft substitutes with enhanced mechanical and biodegradable properties, and biointeractive nature [15,16]. Scaffolds fabricated from (PLLA)/dicalcium phosphate dihydrate (DCPD) composite by indirect casting had also shown to effectively improve the mechanical strength and biocompatibility for the repair of bone defects when compared to the scaffold from neat PLLA [17]. From which the hybrid material may be a better candidate as the cement anti-leakage membrane design [17–19].

In this study, the hybrid membrane made of PLLA and tricalcium silicate (C_3S) was designed to block cement egress during kyphoplasty procedures, with its optimized composition design, mechanical properties, biodegradability, biocompatibility, and anti-leakage capability investigated. This membrane could then find immediate use during kyphoplasty process where vertebral body defects may cause cement leakage.

2. Material and Methods

In this study, tricalcium silicate (C_3S) powders prepared by a sol-gel process were added into poly(L-lactic acid) (PLLA) solution at different ratios to make the composite films using a solution casting method, with their mechanical properties including Young's modulus, maximum stress, and fractural strain measured. Following which the membrane with optimized PLLA/C_3S composition was characterized for its biodegradability, in vivo and in vitro biocompatibility and cytocompatibility, and in vitro anti-leakage capability.

2.1. Preparation of Tricalcium Silicate (C_3S) Powders

Tricalcium silicate powders were synthesized by the sol-gel method, using $Ca(NO_3)_2 \cdot 4H_2O$ and $Si(OC_2H_5)_4$ (TEOS) from Showa, Tokyo, Japan, as the raw materials and nitric acid as a catalyst with an initial CaO/SiO_2 molar ratio of 3. Briefly, 0.5 mol TEOS was added in 200 mL water under continuous stirring. The required amount of $Ca(NO_3)_2 \cdot 4H_2O$, as the calcium precursor, was added to the solution, and the solution was stirred for one hour. Then, the solution was maintained at 80 °C for 24 h until gelation occurred. The gel was dried at 120 °C and then calcined at different temperatures between 1300 and 1400 °C. The resultant powders were ground and sieved to 400-mesh for further characterizations.

The as-prepared Ca_3SiO_5 (C_3S) powders were investigated by X-ray diffraction, XRD (X'Pert[3] Powder, Malvern Panalytical LTD., Malvern, United Kingdom) and scanning electron microscopy, SEM (S-3000H, Hitachi, Berkshire, United Kingdom).

2.2. Fabrication of PLLA/C_3S Composite Membrane

PLLA/C_3S composite membranes used in this study were prepared by the solvent casting technique. The poly(L-lactide) (PLLA, inherent viscosity 0.55–0.75 dL/g in $CHCl_3$) resin from Lactel Absorbable Polymers (Durect Corporation, Birmingham, AL, USA) was preconditioned in drying oven at 40 °C for 48 h to reduce the moisture content before use. The weight ratio of PLLA to C_3S powders from the previous step was 100:1 (wt/%) in 30 mL of chloroform solution. The solution was sealed and stirred for 1 h at room temperature, following which the solution was cast onto a 120 mm × 20 mm (diameter × height) glass plate. The composite films were dried at room temperature in the venting hood for 4 h and then peeled from the glass plate within DI water for 10 min.

2.3. Characterization of PLLA/C_3S Composite Membrane

2.3.1. Tensile Tests

Mechanical properties of composite membranes were performed at room temperature using a universal tensile tester (CY-6102, Chun Yen Testing Machines Co., LTD., Taichung, Taiwan) at a crosshead speed of 10 mm/min. The Young's modulus and maximum tensile strength of the PLLA/C_3S composite membrane were determined from the linear region of stress-strain curves. Five samples of each composition were tested, and the average values ± standard deviation were reported.

2.3.2. Morphological Features

The PLLA/C_3S composite membrane was examined in an SEM (S-3000H, Hitachi, Berkshire, United Kingdom), with the morphology imaged at five keV after mounting and sputter coating with Au.

2.3.3. Contact Angle Measurement

The apparent water-in-air contact angles of the PLLA/C_3S films were measured by the sessile drop method using an FTA1000 goniometer (First Ten Angstroms, Inc., Portsmouth, VA, USA) at room temperature; 2 µL of deionized water (DI water) was dropped on the sample surfaces for the measurement.

2.3.4. The Composite Membrane Porosity

The porosity of membranes under-analyzed was characterized using Mercury Porosimetry (Porous Materials Inc., Ithaca, NY, USA). Mercury porosimetry analysis is based on the intrusion of mercury into the porous membrane structure under controlled pressure applied. The pressure needed for intrusion according to the Washburn-equation [20].

$$D = (1/P) \times 4\gamma \times \cos\theta \quad (1)$$

Where D is the pore diameter, P is the applied pressure, γ is the surface tension of mercury, and θ is the contact angle between the mercury and the membrane sample. The volume of mercury penetrating the pores is measured directly at applied pressure. Membrane porosity refers to the void volume fraction of the membrane and defined as the volume of the pores divided by the total weight of the membrane in this study.

2.3.5. Swelling and Degradation Studies

Swelling and degradation studies of the composite membrane were conducted in DI water under room temperature and phosphate-buffered saline (1× PBS, pH = 7.4, Sigma-Aldrich, Saint Louis, USA) under 37 °C, respectively. Briefly, 0.01 g of oven-dried material was weighed to obtain the dry weight (W_d) and soaked thereafter in 1.0 mL of DI water or PBS for swelling and degradation tests, respectively.

For the swelling test, at each specific time point (i.e., 1, 7, 14 and 21 days), the membrane was removed, blotted gently with filter paper to remove surface water, and the swollen membrane was weighed again (W_w), with the percentage of DI water absorbed by the membrane samples (% WA) calculated using the formula [21]:

$$\% \text{WA} = (W_w - W_d)/W_d \times 100\% \qquad (2)$$

For the degradation test, the degradation percentage was calculated using the following formula [22]:

$$\% \text{WL} = (W_d - W_f)/W_d \times 100\% \qquad (3)$$

Where W_f is the measured weight of the oven-dried membrane samples after soaking in 1.0 mL of PBS in thermostat shock sink at 37 °C, 50 rpm for different days (i.e., 1, 7, 14, 21, and 42 d). All experiments were done in triplicate, and the degradation results were presented as weight loss % (% WL) in the following sections.

2.3.6. Thermal Properties

The glass transition temperature (T_g) and melting temperature (T_m) of the composite was determined by means of a differential scanning calorimeter (DSC 200, NETZSCH, Burlington, MA, US), fitted with a standard DSC cell, and equipped with a Discovery Refrigerated Cooling System (RCS90) (all TA Instruments, Delaware, USA). Samples of about 7 mg were placed into aluminum pans and subjected to two heating cycles from −90 °C to +110 °C with cooling and heating rates of 10 °C/min. The DSC cell was purged with dry nitrogen at 50 mL/min. The system was calibrated both in temperature and enthalpy with an Indium standard. Results are given as average value. All data were processed with Thermal Analysis software (TA Instruments, Delaware, USA).

2.4. Biocompatibility and Cytotoxicity of Membranes

2.4.1. MTT Assay

In order to evaluate the cytotoxicity of composite membranes, extracts of the material and MTT (3-dimethylthiazol-2,5-diphenyltetrazolium bromide; Aldrich 135038, Sigma-Aldrich, Saint Louis, USA) assay were applied using L929 mouse fibroblasts as stated in the International Organization for Standardization 109993-5 [23,24]. Briefly, the extracts were prepared with sterile membrane samples incubated in the cell-cultured medium (1 g membrane in 5 mL of medium) at 37 °C under 5% CO_2 atmosphere and extracted after 1, 7, 14, and 21 days. L929 fibroblasts were cultured in 24-well plates at a density of 5×10^4 cell/well for 24 hr, with 1 mL of Dulbecco's modified Eagle's medium (DMEM, Gibco, Dublin, Ireland), containing 1% Non-essential amino acid (NEAA, Gibco), 1% Sodium pyruvate (SP, Gibco), 1% L-glutamine (L-G, Gibco), 1% prostate-specific antigen (PSA, Gibco) and 10% fetal bovine serum (FBS, Gibco). After that, the medium was replaced with 1, 7, 14, and 21-day extracts. After another 24 h, the extracts were removed, with 0.5 mL of 0.5 mg/mL MTT solution (0.2 mL of MTT mother solution (5 mg/mL of MTT in phosphate-buffered saline (1× PBS, pH = 7.4)) and 1.8 mL DMEM) added to each cell-containing well and incubated for about 4 h at 37 °C. Then, the MTT solution was removed carefully, and 1 mL of DMSO (D2650, Sigma, Saint Louis, USA) was added to each well to dissolve the formed formazan purple crystals. The formation of formazan product was analyzed by measuring absorbance at 570 nm using an ELISA microplate reader. The mitochondrial activity of the L929 fibroblasts was expressed as the result of cell viability (%): Relative absorbance from each condition over the control cultures (i.e., cells seeded on microplate and cultured in normal medium). Three replicates were analyzed for each sample, and the final data were expressed as the mean value ± the standard deviation.

2.4.2. MG-63 Culture on Composite Membrane

Human MG-63 osteoblast-like cells were purchased from Bioresource Collection and Research Center (product No. BCRC 60279, passage 101, BCRC, Hsinchu City, Taiwan). A membrane sample such as PLLA or PLLA/C_3S composite (10 mm × 10 mm) was placed in a 48-well plate (Thermo Fisher Scientific Inc., USA). After which, the MG-63 cells were seeded at a density of 5.0×10^4 cells/well onto the plastic well of the plate in presence of the membrane and grow to confluence with 1 mL of DMEM (Gibco) containing 1% NEAA, 1% SP, 1% L-G, 1% PSA, and 10% FBS in each well. After 7, 14, and 7 days of culture, the medium in each well was removed and each well was rinsed with phosphate-buffered saline (1× PBS, pH = 7.4) three times with the remaining cells for the following investigations.

2.4.3. Alkaline Phosphatase Activity

ALP activity measurement was carried out to analyze the effect of phosphorylated chito-oligosaccharides on the osteoblast-like MG-63 cell line. Following the previous steps, the cells in each well were homogenized with 0.5 mL of 0.1% Triton-X100 in a sonicator for 5 min. The cellular activity was then measured by incubating for 60 min at 37 °C in 250 mM carbonate buffer containing 1.5 mM Magnesium chloride hexahydrate ($MgCl_2$, Sigma, Saint Louis, USA) and 15 mM para-Nitro Phenyl Phosphate (p-NPP, J.T. Baker, Radnor, PA, USA). In the presence of ALP, p-NPP is transformed to p-nitro phenol and inorganic phosphate. The ALP activity was determined by measuring the absorbance at 405 nm in a spectrophotometer [25].

2.4.4. In Vivo Animal Test and Surgical Procedures

In vivo animal studies were performed to assess the biocompatibility of the PLLA/C_3S composites. All chemicals in this section were purchased from Sigma-Aldrich (Saint Louis, USA) and used without additional purification unless otherwise specified. All experimental animal protocols were carried out according to the Guide for the Care and Use of Laboratory Animals and approved by the National Taipei University of Technology, Taipei, Taiwan. All procedures were performed in Chang Gung Memorial Hospital with prior approval of the Institutional Animal Care and Use Committee (IACUC) of Chang Gung Memorial Hospital, Taoyuan City, Taiwan (2016112801). The animals used in this study were male Sprague-Dawley (SD) rats of 300 g body weight, on average, purchased from the National Laboratory Animal Center of Taiwan and acclimatized for a minimum of one week before experimentation. Animals had ad libitum access to standard rat chow and water at all times. All experimental procedures on given animals were completed under aseptic conditions. Each animal was anesthetized with 0.3 mL of a 1:2 mixture of Zoletil and Rompun. Once the rats were anesthetized, the dorsal hair was removed and cleaned using 70% volume fraction ethanol before the surgical procedures, as listed in Figure 1(a). Four horizontal incisions approximately 15 mm in length were made along each side on the back of the rat, and subcutaneous skin pockets were created by blunt dissection (Figure 1(b)). The pockets were separated by 40 mm to 50 mm. Each cylindrical-shaped sample (4.5 mm in diameter and 20 mm in length) was inserted into a pocket of subcutaneous tissues as shown in Figure 1(b), with 4 different samples (sample A (PMMA), B (PLLA), C (PLLA/C_3S), and D (PLLA/C_3S tube filled with PMMA)) implanted and located at different locations as shown in Figure 1(c). Subsequently, the incision was properly sutured, cleaned with 2% iodine, and dressed. For in vivo experiments, all implantations were one-shot treatments, with no replacement at any time point. The tissue specimens were obtained from four male SD rats after treatment for 7, 14, 21, and 28 days, with the portion of the selected tissue fixed with paraformaldehyde for 24 h at 4 °C, frozen and sectioned for hematoxylin-eosin (H&E) staining. The histological sections were examined using a stereomicroscope (Stereo Discovery, V20; Zeiss, Germany). All histological analyses were performed with at least three wounds per group per time point.

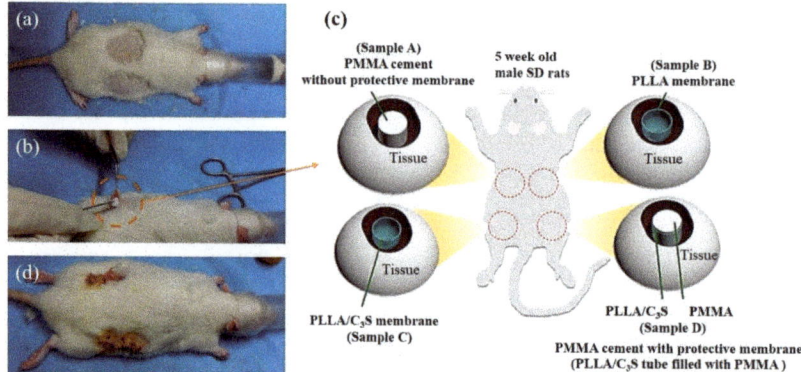

Figure 1. Implantation workflow. (**a**) Mice dorsal hair was removed and the skin was disinfected, (**b**) incision and the subcutaneous pocket on the rat back is formed with four different materials implanted with each of the material explained in (**c**), including polymethyl methacrylate (PMMA) (samples A), poly(L-lactic acid) (PLLA) membrane (sample B), PLLA/C_3S membrane (sample C), and PLLA/C_3S membrane tube filled with PMMA cement (sample D) prepared in a cylinder shape with the dimension of 20mm × 4.5mm (height × diameter). Sample A was considered as the positive control compared to other groups in this study. After which, (**d**) the incision was properly sutured and dressed.

2.5. Anti-Leakage Tests of the Composite Membrane

The in vitro test of composite membrane for preventing bone cement leakage was conducted using fresh lumbar vertebral bones dissected from pigs and randomly divided into the blank group (without membranes), control group (with PLLA membrane), and the experimental group (with PLLA/C_3S). The extra muscles, ligaments, and spinal marrow were eliminated and the vertebrae with the upper and lower endplates were completely retained. The models of vertebral compression fractures with fissures in anterior vertebrae were prepared by drilling two holes to a depth of 20 mm with a diameter of 4.5 mm. A small device called a balloon-tamp is covered with the composite membrane and then inserted through the needle and into the fractured vertebra. When the balloon-tamp is removed, it leaves a cavity covered with the membrane firstly and filled with the PMMA bone cement and then immediately sent for anti-leakage investigation using micro-computed tomography (Micro-CT).

All vertebrae samples were imaged using a benchtop Micro-CT imager (SkyScan 1076; Bruker-MicroCT, Kontich, Belgium) at 35 μm voxel image resolution with 100 kV, 100 μA, and a 1.0 mm aluminum filter.

2.6. Statistical Analysis

The results are expressed as the mean ± standard deviation (SD) with N ≧ 3. Data were analyzed by one-way ANOVA with the statistical function of Microsoft Excel.

3. Results and Discussions

In this context, we fabricated an implantable PLLA/C_3S composite membrane that might be able to prevent the leakage of PMMA cement after kyphoplasty procedure, with its mechanical, in vivo, and in vitro biocompatibility, cytocompatibility, and in-vitro anti-leakage properties investigated.

3.1. Synthesis of Ca_3SiO_5 Powders

PLLA/C_3S composite membranes used in this study were prepared by the solvent casting technique, with tricalcium silicate powders synthesized by the sol-gel method [26] in our lab using $Ca(NO_3)_2·4H_2O$ and $Si(OC_2H_5)_4$ (TEOS) as the raw materials. Figure 2(a) shows the XRD pattern of prepared Ca_3SiO_5 powder, also named C_3S, with the results showing that it is triclinic and is the NO. 49-0442 in PDF standard card [27]. The grain size and surface morphologies of Ca_3SiO_5 powders were shown by the SEM micrographs (Figure 2(b)) which indicated that the Ca_3SiO_5 powders have a polygonal shape with the particle size of about 10–30 μm.

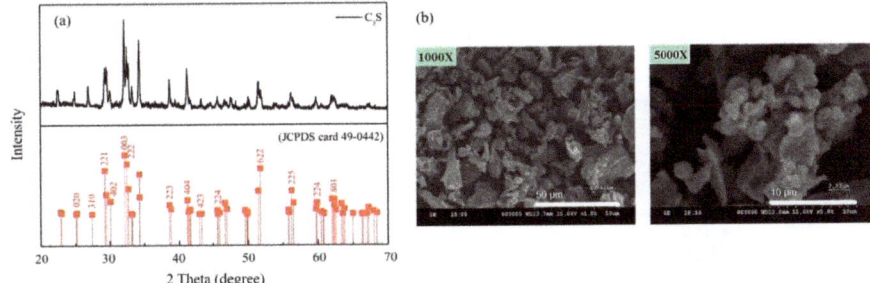

Figure 2. (a) XRD pattern and (b) SEM micrograph of prepared C_3S powders.

3.2. Characterization of PLLA/C_3S Composite Membrane

The hybrid membrane made of PLLA and tricalcium silicate (C_3S) was fabricated with its characterizations reported and discussed in the following sections.

3.2.1. Tensile Properties

As shown in Figure 3, the strain at the break (Figure 3(a)) and maximum tensile stress (Figure 3(b)) of PLLA/C_3S composite membrane was reported based on the varying weight percent of C_3S in PLLA (i.e., 0%, 0.1%, 0.2%, 0.5%, 1%, 1.5%, and 2% (wt %). Accordingly, the maximum strength of the composite membrane increased with increasing C_3S concentration, while the strain at break decreased when the C_3S content is higher. The increased concentration of C_3S might act as reinforcing filler for the increased bonding conditions of PLLA/C_3S composite [28], which leads to its higher tensile strength and stronger resistance to the shape changes. From which the durability of the composite membrane, which shows the ability of the membrane specimen to resist changes of shape without cracking was decreased with the increasing concentration of C_3S as well [29], which is in agreement with the general trend observed in Figure 3(a). However, our material design was aimed to cover the gap with the irregular shape prepared in a fractured vertebral body during the kyphoplasty in which case higher durability of the composite membrane is required. Therefore, to keep the relative higher tensile strength for a better barrier to prevent the cement egression while keeping higher plasticity or strain at break of our composite membrane design, 1% (wt %) of C_3S in PLLA was selected accordingly as the optimized PLLA/C_3S ratio for further membrane characterizations with the results discussed in the following sections.

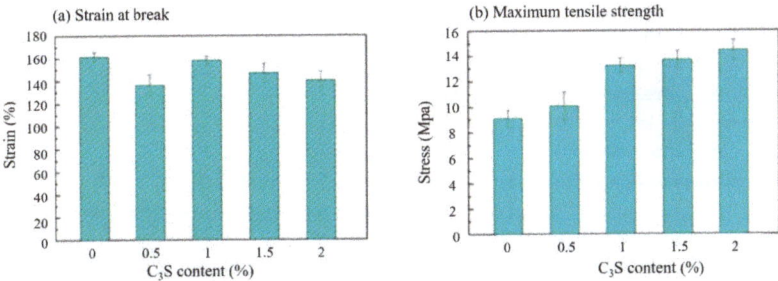

Figure 3. Tensile properties of PLLA/C_3S composite membranes prepared at a various weight percent of C_3S in PLLA (wt %) by (a) strain and (b) stress.

3.2.2. SEM Morphology and Surface Distribution of C_3S within the Composite Membrane

Figure 4 showed SEM pictures of C_3S particles precipitated in a PLLA membrane (1% (w/w)), with the membrane thickness presented from the cross-sectional SEM micrograph in Figure 4(c) and 4(f) for PLLA and PLLA/C_3S membranes, respectively. The distribution of C_3S within the membrane was further investigated using X-ray fluorescence (XRF) spectroscopy to characterize the chemical composition of Ca and Si on our composite membranes coated on a copper substrate. The precipitations of C_3S on our membrane is significant for the properties of the hybrid biomaterials due to the enhanced binding affinity between the PLLA structures from the uniformly distributed Ca and Si within the membrane. As shown, the raw elemental maps for Ca and Si within the membrane were given in Figure 4(g) and 4(h), respectively, which indicated the uniform distribution of C_3S content throughout the PLLA/C_3S membrane.

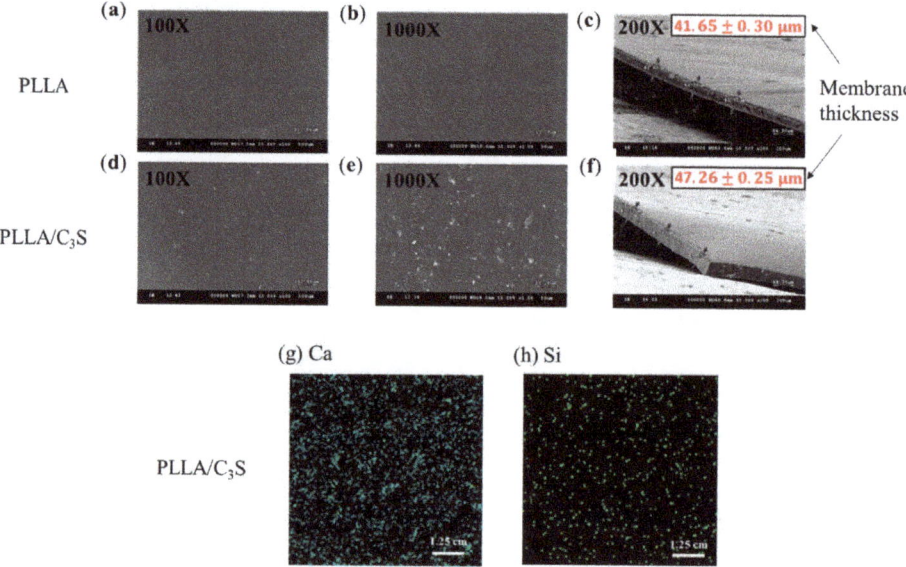

Figure 4. SEM micrograph of (a,b) prepared PLLA and (d,e) PLLA/C_3S membranes. Membrane thickness was presented as mean ± SD with N = 3, as shown in the cross-sectional SEM micrograph of 4(c) and 4(f) for PLLA and PLLA/C_3S membranes respectively. XRF mappings of chemical compositions of (g) Ca and (h) Si were shown within the prepared PLLA/C_3S membranes.

3.2.3. Contact Angle Measurement

Contact angle measurement is a standard method used to determine the hydrophilicity of a biomaterial surface. Figure 5 presents the contact angle values of PLLA and PLLA/C_3S (i.e., 1% (w/w) of C_3S in PLLA) composite determined by static contact angle measurement. Pure PLLA exhibits a contact angle of 76.3°, in agreement with the hydrophobic nature of the polymer [30]. With the addition of the C_3S content (i.e., 1% (w/w) of C_3S in PLLA), the contact angle decreases to 68.5° for PLLA/C_3S composite, probably due to a reduction of interfacial tension between composite and water from the hydrophilic nature of C_3S, making the PLLA/C_3S composite surface more hydrophilic. Although the cell behaviors were found to be only partially influenced by the implanted biomaterial surfaces, cell spreading on the hydrophilic surface was usually more enhanced and uniformly spread compared to cells spreading on hydrophobic surfaces, due to a more complicated protein adsorption behavior involved such as unfolding on the hydrophobic substrate [31]. In which case PLLA/C_3S could be a better-implanted material regarding biological applications when compared to the usage of neat PLLA polymer.

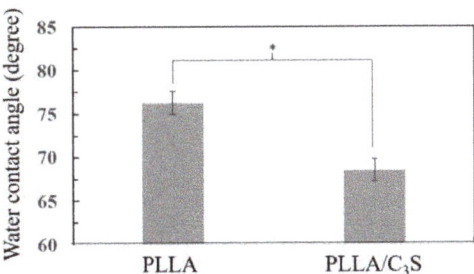

Figure 5. The water contact angle of PLLA and PLLA/C$_3$S membrane, with data presented as mean ± SD for N > 3. (*P < 0.05).

3.2.4. Porosimetry Measurement

As shown in Figure 6(a) and (b), although both PLLA and PLLA/C$_3$S (i.e., 1% (w/w) of C$_3$S in PLLA) composite membrane have similar pore diameter in average, ranging from 20 to 30 nm, the broader size distribution of PLLA matrix was observed when compared to that of PLLA/C$_3$S. This was further explained by the total intrusion volume data in Figure 6(c), also called the interstitial void volume (mL/g), representing the space between packed particles [32], with the interstitial void volume of PLLA matrix decreased from 38.8 to 22.9 (mL/g) after an addition of C$_3$S (1% (w/w) of C$_3$S in PLLA), which is also in agreement with our previous findings that C$_3$S particles added to increase the binding conditions within PLLA matrix, which makes the tighter packing of PLLA chains as well as the narrower distributions of particle sizes.

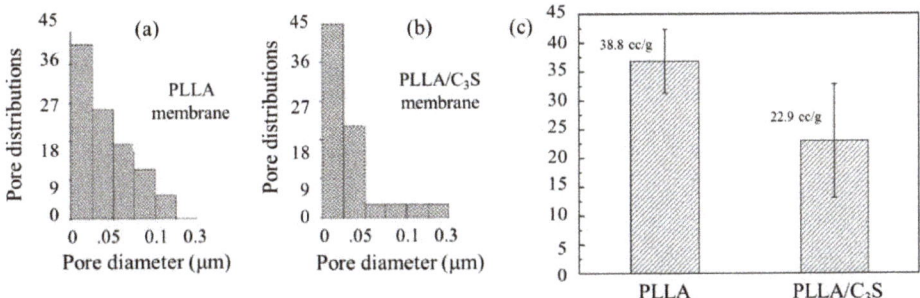

Figure 6. Porosity measurement of PLLA and PLLA/C$_3$S membrane. Pore size and distributions of (a) PLLA and (b) PLLA/C$_3$S. Total intrusion volumes of both membranes were presented in (c).

3.2.5. Swelling and Degradation Properties

Poly(L-lactide) (PLLA) is more resistant to the degradation when compared to PLA alone due to its higher crystallinity, which makes the tensile strength of PLLA last longer [33]. The crystallinity of a polymer is also related to its swelling properties, which were primarily involved with the molecular chains in the amorphous region between lamella crystals, with more of these chains (e.g., higher amorphous contents or lower crystallinity) involved leading to a faster degradation rate due to the preferentially hydrolyzed properties of these chains [34,35]. As shown in Figure 7(a), the swelling index of PLLA has no significant difference when compared to that of PLLA/C$_3$S composite, which suggests that the addition of C$_3$S did not compromise the crystallinity of PLLA. This was further supported by the similar degradation profiles observed between PLLA and PLLA/C$_3$S composite, as shown in Figure 7(b).

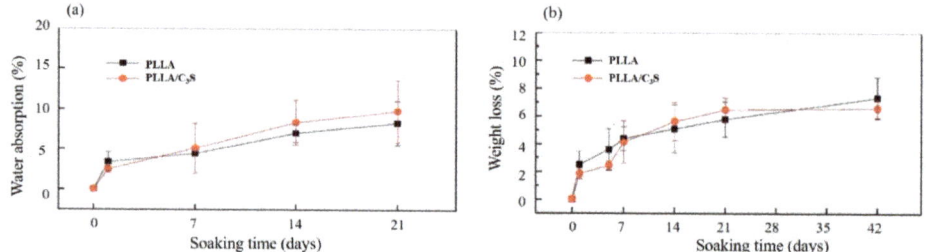

Figure 7. Swelling properties and degradation profiles of (**a**) PLLA and (**b**) PLLA/C$_3$S membrane in DI water at room temperature and in phosphate-buffered saline (1× PBS, pH = 7.4) under 37 °C, respectively. Data represent mean ± SD with N = 3.

3.2.6. Thermal Properties

DSC produces a heat flow (W/g) versus temperature (Celsius) curve, with the area under a melting transition curve of DSC trace representing the total amount of heat absorbed during the sample melting process, which would be proportional to the percentage crystalline by weight of the samples [36]. Figure 8 shows the melting endotherm for PLLA and PLLA/C$_3$S composite samples during the heating process. As shown, these similar melting profiles suggest the crystallinity between PLLA and PLLA/C$_3$S composite are not significantly different, which also confirms the intact PLLA crystallinity after the addition of 1% (*w/w*) of C$_3$S.

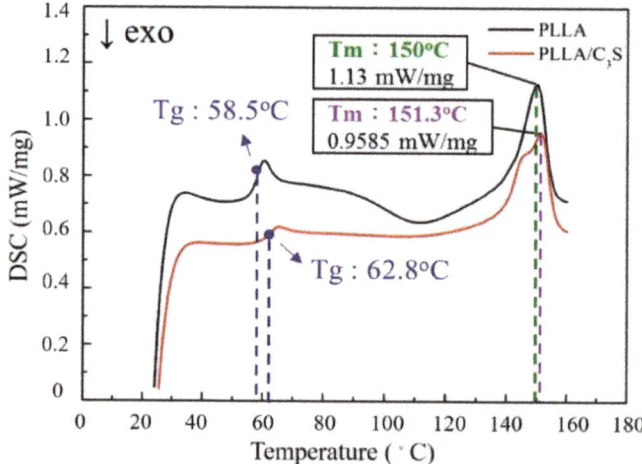

Figure 8. DSC scans of PLLA and PLLA/C$_3$S composite melting thermograms.

3.3. Cellular Response

Cytocompatibility of PLLA and PLLA/C$_3$S composite membranes were investigated using a MTT metabolic activity assay and the activity of alkaline phosphatase measurement, to observe the cell viability as well as cellular differentiation and proliferation when cultured with the PLLA and PLLA/C$_3$S composite membranes.

3.3.1. MTT Assay

This test was performed according to the International Organization for Standardization (ISO) norms [24], using the indirect contact method with extracts as applied to L929 cells, a line derived from mouse fibroblasts [23]. A MTT cell viability assay was performed to evaluate the effect of the C_3S addition on the cellular performance of the PLLA/C_3S composite samples in terms of cellular viability. As shown in Figure 9, the MTT results of the cytotoxicity levels of all sample extracts (i.e., blank group, PLLA, and PLLA/C_3S composite) in direct contact with the L929s were given. As mentioned earlier, sample extracts were prepared with sterile samples such as membranes incubated in the cell-cultured medium (1 g sample in 5 mL of medium) and extracted after 7, 14, and 21 d. Cells grown within pure medium were used as blanks. Cell viability in these extracts was based on the mitochondrial enzymatic conversion of the tetrazolium into MTT formazan precipitated, occurring only in metabolically active cells. The amount of formazan dye directly correlates with the number of viable cells presented in the sample extracts [37]. The results indicated that the PLLA/C_3S specimens had a significantly higher cell growth observed in comparison with other groups at each time points, with $p < 0.05$ indicated by one asterisks (*), $p < 0.01$ indicated by two asterisks (**) and $p < 0.001$ indicated by three asterisks (***). Results confirmed the lack of toxicity for all of the investigated samples.

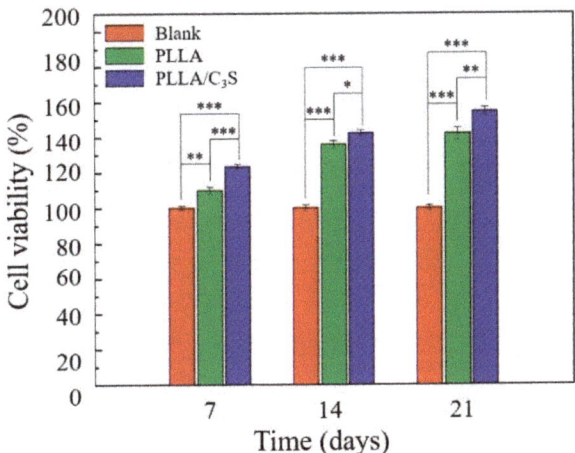

Figure 9. MTT cell viability measurement of L929 cells, which were seeded onto the surface of tissue culture plate (polystyrene) and cultured with PLLA or PLLA/C_3S composite membranes. Cells grown within pure medium were used as blanks. An asterisk (*) indicates a statistically significant difference ($p < 0.05$); two asterisks (**) indicate a significant difference with $p < 0.01$, and three asterisks (***) indicate a significant difference with $p < 0.001$. No material interference. We had the membrane samples incubated in cell culture media without the cells, with the collected media showing no absorbance at 570 nm (data not shown), which confirms that nothing from the sample interfered with the absorbance at 570 nm.

3.3.2. Alkaline Phosphatase (ALP) Activity Measurement

Ceramic compositions have been shown to promote proliferation and differentiation of human osteoblasts into osteogenic lineages [38,39]. How the membrane material applied in this study will affect the cellular differentiation will be determined by ALP as an early marker of osteogenic differentiation, with the ALP playing a major role in the formation of mineral deposits in the matrix during new bone formation [40].

Cell differentiation and proliferation of MG-63 induced by PLLA and PLLA/C_3S membranes were evaluated by quantitative measurement of ALP activity. As shown in Figure 10, the ALP activity of the MG-63 cells increased in all groups for the incubation period of up to 21 days. On the 7th day, there were no significant difference observed in the ALP activity of the MG-63 cells cultivated on both membranes. On the 14th day, the ALP activity of MG-63 cells grown on PLLA/C_3S increased significantly compared to that on PLLA ($p < 0.05$). On the 21st day, more significant differences were observed in the ALP activity of MG-63 cells cultured on PLLA/C_3S and PLLA membranes with $p < 0.001$, which indicates that PLLA/C_3S could promote more of the differentiation of MG-63 cells. Therefore, the results of ALP activity showed the significance of C_3S in cell differentiation, where the presence of C_3S in PLLA might cause stimulation of bone cell response.

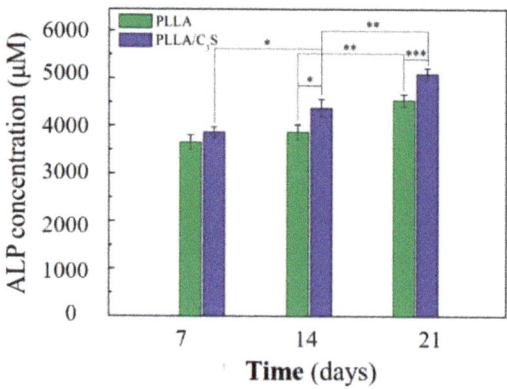

Figure 10. Cellular alkaline phosphatase (ALP) activity of MG-63 on PLLA and PLLA/C_3S membranes. An asterisk (*) indicates a statistically significant difference ($p < 0.05$); two asterisks (**) indicate a significant difference with $p < 0.01$, and three asterisks (***) indicate a significant difference with $p < 0.001$.

3.4. Qualitative Analysis of in Vivo Animal Tests

The purpose of this in vivo test was for the confirmation of the anti-leakages of PMMA cements from PLLA/C_3S membrane and only the qualitative analysis of cellular reaction over time was studied. The PMMA cement, polymeric membranes (i.e., PLLA or PLLA/C_3S membranes), and PMMA cement wrapped with PLLA/C_3S membrane were investigated by implanting them in the subcutaneous pouch created on the dorsum of the Sprague-Dawley (SD) rats. Reentry of the implant site was done at different time points: Day 7, 14, 21 and 28 to assess the impact of polymers implanted on subcutaneous tissue, with the visual inspection showing that, the PMMA cement wrapped with PLLA/C_3S membrane was appearing with the full integrity of the cavity in histologic specimens at all-time points which further suggest the anti-leakages of PMMA cements from PLLA/C_3S membrane.

As shown in Figure 11, seven days after implantation, the membrane maintained its native structure and integrity and induced mononuclear cells, which were found on the membrane surface and the central region of each material was mostly free of cells. Although the presence of lymphocytes in histologic specimens of tissues surrounding implants have been present for several days, which might indicate that the inflammatory response to the implanted material is not over [41], the overall inflammatory response decreases after a longer implantation period. For example, after fourteen days following implantation, the numbers of lymphocytes and macrophages counted in averages/per section infiltrated at the site of implanted materials decreases, with more of the mononuclear cells invading the PLLA membrane observed when compared to that of PLLA/C_3S implant. However, because the lymphocytes did not distribute homogenously, it might not be representative to present the average number of lymphocytes per section, and the quantitative histomorphometrical analysis of our samples is required for clinical practice in future research with the proper sampling method provided [42].

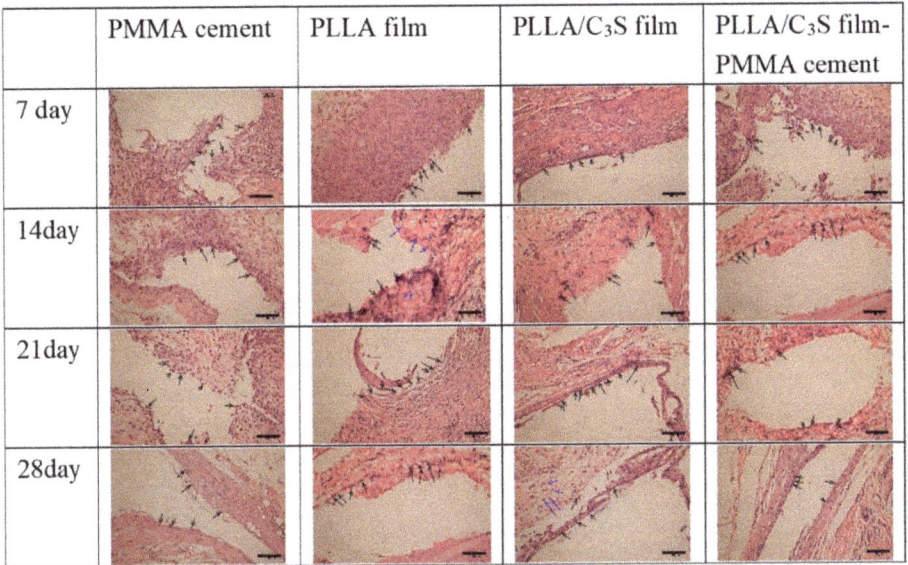

Figure 11. Histology of Sprague-Dawley (SD) rats skin. Each frame, from top to bottom, illustrates the histology of the tissues surrounding samples after different time points of subcutaneous implantation in a rat, with each sample appearing as the cavity surrounded by lymphocytes or macrophages, indicated by the black and blue arrows, respectively. (H&E x20; scale bar = 100 μm).

3.5. Anti-Leakage Tests

This final section was performed as the first investigation to assay the effects of PLLA/C_3S composite on preventing cement leakage, with its in vitro observation of potential anti-leakage properties during the process of kyphoplasty to block cement egress through the posterior wall defect tested using fresh pig lumbar vertebral bones with holes drilled to represent the vertebral compression fractures and covered with the composite membrane using a balloon-tamp (Figure 12(a)). When the balloon-tamp is removed, it leaves a cavity with or without a PLLA or PLLA/C_3S membrane-covered within the vertebral bones, which was then filled with PMMA bone cement soon after for anti-leakage investigation (Figure 12(b)). The photos and micro-computed tomography (Micro-CT) images are shown in Figure 12.

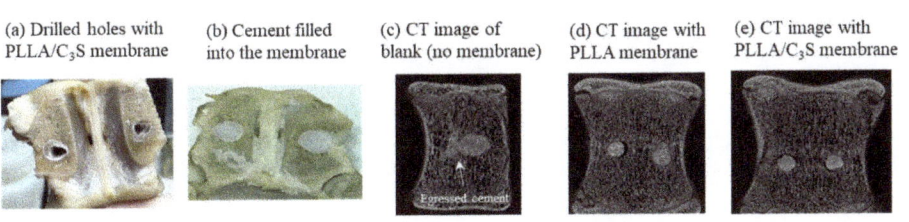

Figure 12. Anti-leakage investigation was carried out on simulated vertebral compression fractures by two drilled holes, in which the cavity within the pig vertebral bones was then filled with PMMA cement with or without membrane covered. (**a**) Photo of drilled holes with PLLA/C_3S membrane; (**b**) photo of PMMA cement filled into the PLLA/C_3S membrane; (**c**) micro-CT image without membrane (blank); (**d**) micro-CT image with PLLA membrane covered; (**e**) micro-CT image with PLLA/C_3S membrane covered.

Our findings showed that, when the PLLA (Figure 12(d)) or PLLA/C$_3$S composite (Figure 12(e)) membranes were applied during the kyphoplasty treatment before the PMMA cement injection, a better barrier outcome for the prevention of cement leakage was observed, especially when compared with the case without membrane used (blank), which has the significant amount of cement egressed through the posterior wall defect observed after the kyphoplasty (Figure 12(c)). Although both membranes, PLLA or PLLA/C$_3$S, might be used as anti-cement leakage purpose, considering the lower tensile strength, bigger pore size, and more hydrophobic features of PLLA membrane, as discussed in our previous sections, the PLLA/C$_3$S composite membrane should be a more reliable barrier with good potential to prevent the cement leakage in selected cases of vertebrae with posterior wall defects after the kyphoplasty procedure.

Of course, limitations of this experimental study at this stage merit to be mentioned. First of all, this is an in vitro study conducted in artificial vertebral analogs under selected environmental conditions (such as the fixed temperature and time of injections), which might be significantly varied in the clinical practice, with the cement leakage patterns much less predictable. Secondly, PMMA cement was the only cement tested using this anti-leakage membrane, which may not be applicable when other cements are applied. Many investigations have shown that cement leakage patterns in a vertebral body model may vary significantly especially when the viscosity of the cement was changed during the kyphoplasty procedures [43,44]. These weaknesses that may become obvious in clinical practice will be addressed in our continuous testing of PLLA/C$_3$S composite membrane in cement augmentation techniques in selected cases of vertebrae with posterior wall defects. Finally, the original design of percutaneous kyphoplasty (PKP) has been developed as a treatment in which a gap was prepared in a fractured vertebral body using a balloon, followed by an injection of bone cement to restore the vertebral height, but the molecular mechanisms underlying such an effect long remained undetermined. For example, in previous histological observations, the PKP might evoke a local environment in the defect which might be conducive to the molecular mechanisms underlying coupled bone formation and bone remodeling during fracture healing via the recruitment and differentiation of multiple cell types, including osteoblasts and osteoclasts [45–47]. From this, the biological events in both the membrane and the underlying defect are important for bone regeneration and must be a key for a better design of this hybrid membrane made of PLLA and tricalcium silicate (C$_3$S) to block cement egress during kyphoplasty procedures, which should be discussed in our future studies.

4. Conclusions

We have successfully created a potential solution for the posterior leakage during the kyphoplasty procedures using PLLA/C$_3$S composite as anti-leakage membrane, with the addition of tricalcium silicate (C$_3$S) into PLLA scaffold showing enhanced mechanical and anti-degradation properties while keeping good cytocompatibility, when compared to PLLA alone. Most importantly, when this material design was applied under standardized PMMA cement injection conditions, no posterior wall leakage was observed at the standard kyphoplasty procedure using pig lumbar vertebral bone as test models. In which case, the PLLA/C$_3$S composite membrane might be a useful barrier material for implantable balloon kyphoplasty.

Author Contributions: Conceptualization, J.-T.H. and R.-J.C.; Data curation, Y.W., Y.-S.H. and X.-J.H.; Formal analysis, X.-J.H.; Funding acquisition, T.-H.L., J.-T.H. and R.-J.C.; Investigation, T.-H.L., Y.-S.H. and X.-J.H.; Methodology, J.-T.H. and R.-J.C.; Project administration, R.-J.C.; Supervision, T.-H.L. and J.-T.H.; Writing—original draft, Y.W.; Writing—review & editing, T.-H.L. and R.-J.C.

Funding: This research was funded by Ministry of Science and Technology, Taiwan: MOST 104-2221-E-027-061 MOST 105-2221-E-038-011, MOST 106-2221-E-038-009 and MOST 106-2221-E-027-034.

Acknowledgments: The authors are grateful for the financial support to this research by the Ministry of Science and Technology of Taiwan, under Grant No. MOST 104-2221-E-027-061 MOST 105-2221-E-038-011, MOST 106-2221-E-038-009 and MOST 106-2221-E-027-034. The first and second authors contributed equally to this work.

Conflicts of Interest: The authors declare no conflicts of interest.

References

1. Sözen, T.; Özışık, L.; Başaran, N.Ç. An overview and management of osteoporosis. *Eur. J. Rheumatol.* **2017**, *4*, 46–56. [CrossRef] [PubMed]
2. Jung, H.J.; Park, Y.-S.; Seo, H.-Y.; Lee, J.-C.; An, K.-C.; Kim, J.-H.; Shin, B.-J.; Kang, T.W.; Park, S.Y. Quality of Life in Patients with Osteoporotic Vertebral Compression Fractures. *J. Bone Metab.* **2017**, *24*, 187–196. [CrossRef] [PubMed]
3. Hsieh, M.K.; Chen, L.H.; Chen, W.J. Current concepts of percutaneous balloon kyphoplasty for the treatment of osteoporotic vertebral compression fractures: Evidence-based review. *Biomed. J.* **2013**, *36*, 154–161. [CrossRef] [PubMed]
4. Huang, Y.-S.; Hao, D.-J.; Feng, H.; Zhang, H.-P.; He, S.-M.; Ge, C.-Y.; Niu, X.-B. Comparison of Percutaneous Kyphoplasty and Bone Cement-Augmented Short-Segment Pedicle Screw Fixation for Management of Kümmell Disease. *Med. Sci. Monit. Int. Med. J. Exp. Clin. Res.* **2018**, *24*, 1072–1079. [CrossRef] [PubMed]
5. Zhan, Y.; Jiang, J.; Liao, H.; Tan, H.; Yang, K. Risk Factors for Cement Leakage After Vertebroplasty or Kyphoplasty: A Meta-Analysis of Published Evidence. *World Neurosurg.* **2017**, *101*, 633–642. [CrossRef]
6. Liu, T.; Li, Z.; Su, Q.; Hai, Y. Cement leakage in osteoporotic vertebral compression fractures with cortical defect using high-viscosity bone cement during unilateral percutaneous kyphoplasty surgery. *Medicine* **2017**, *96*, e7216. [CrossRef]
7. Lai, P.-L.; Tai, C.-L.; Chen, L.-H.; Nien, N.-Y. Cement leakage causes potential thermal injury in vertebroplasty. *BMC Musculoskelet. Disord.* **2011**, *12*, 116. [CrossRef]
8. Anselmetti, G.C.; Zoarski, G.; Manca, A.; Masala, S.; Eminefendic, H.; Russo, F.; Regge, D. Percutaneous vertebroplasty and bone cement leakage: Clinical experience with a new high-viscosity bone cement and delivery system for vertebral augmentation in benign and malignant compression fractures. *Cardiovasc. Interv. Radiol.* **2008**, *31*, 937–947. [CrossRef]
9. Shridhar, P.; Chen, Y.; Khalil, R.; Plakseychuk, A.; Cho, S.K.; Tillman, B.; Kumta, P.N.; Chun, Y. A Review of PMMA Bone Cement and Intra-Cardiac Embolism. *Materials* **2016**, *9*, 821. [CrossRef]
10. Greene, D.L.; Isaac, R.; Neuwirth, M.; Bitan, F.D. The eggshell technique for prevention of cement leakage during kyphoplasty. *J. Spinal Disord. Tech.* **2007**, *20*, 229–232. [CrossRef]
11. DalCanto, R.A.; Reinhardt, M.K.; Lieberman, I.H. Double cement-application cavity containment kyphoplasty: Technique description and efficacy. *Am. J. Orthop. Belle Mead N.J.* **2009**, *38*, E110–E114.
12. Inoue, M.; Sakane, M.; Taguchi, T. Fabrication of reactive poly(vinyl alcohol) membranes for prevention of bone cement leakage. *J. Biomed. Mater. Res. Part B Appl. Biomater.* **2014**, *102*, 1786–1791. [CrossRef] [PubMed]
13. Amini, A.R.; Laurencin, C.T.; Nukavarapu, S.P. Bone tissue engineering: Recent advances and challenges. *Crit. Rev. Biomed. Eng.* **2012**, *40*, 363–408. [CrossRef] [PubMed]
14. Costa, V.C.; Costa, H.S.; Vasconcelos, W.L.; Pereira, M.d.M.; Oréfice, R.L.; Mansur, H.S. Preparation of hybrid biomaterials for bone tissue engineering. *Mater. Res.* **2007**, *10*, 21–26. [CrossRef]
15. Tatullo, M.; Spagnuolo, G.; Codispoti, B.; Zamparini, F.; Zhang, A.; Esposti, M.D.; Aparicio, C.; Rengo, C.; Nuzzolese, M.; Manzoli, L.; et al. PLA-Based Mineral-Doped Scaffolds Seeded with Human Periapical Cyst-Derived MSCs: A Promising Tool for Regenerative Healing in Dentistry. *Materials* **2019**, *12*, 597. [CrossRef] [PubMed]
16. Gandolfi, M.G.; Zamparini, F.; Degli Esposti, M.; Chiellini, F.; Aparicio, C.; Fava, F.; Fabbri, P.; Taddei, P.; Prati, C. Polylactic acid-based porous scaffolds doped with calcium silicate and dicalcium phosphate dihydrate designed for biomedical application. *Mater. Sci. Eng. C Mater. Biol. Appl.* **2018**, *82*, 163–181. [CrossRef] [PubMed]
17. Tanataweethum, N.; Liu, W.C.; Goebel, W.S.; Li, D.; Chu, T.M. Fabrication of Poly-l-lactic Acid/Dicalcium Phosphate Dihydrate Composite Scaffolds with High Mechanical Strength-Implications for Bone Tissue Engineering. *J. Funct. Biomater.* **2015**, *6*, 1036–1053. [CrossRef]
18. Roeder, R.K.; Converse, G.L.; Kane, R.J.; Yue, W. Hydroxyapatite-reinforced polymer biocomposites for synthetic bone substitutes. *Jom* **2008**, *60*, 38–45. [CrossRef]
19. Lv, Y.; Li, A.; Zhou, F.; Pan, X.; Liang, F.; Qu, X.; Qiu, D.; Yang, Z. A Novel Composite PMMA-based Bone Cement with Reduced Potential for Thermal Necrosis. *ACS Appl. Mater. Interfaces* **2015**, *7*, 11280–11285. [CrossRef]

20. Vitas, S.; Segmehl, J.S.; Burgert, I.; Cabane, E. Porosity and Pore Size Distribution of Native and Delignified Beech Wood Determined by Mercury Intrusion Porosimetry. *Materials* **2019**, *12*, 416. [CrossRef]
21. Guo, Z.; Yang, C.; Zhou, Z.; Chen, S.; Li, F. Characterization of biodegradable poly(lactic acid) porous scaffolds prepared using selective enzymatic degradation for tissue engineering. *RSC Adv.* **2017**, *7*, 34063–34070. [CrossRef]
22. Spagnuolo, M.; Liu, L. Fabrication and Degradation of Electrospun Scaffolds from L-Tyrosine-Based Polyurethane Blends for Tissue Engineering Applications. *ISRN Nanotechnol.* **2012**, *2012*, 11. [CrossRef]
23. Cannella, V.; Altomare, R.; Chiaramonte, G.; Di Bella, S.; Mira, F.; Russotto, L.; Pisano, P.; Guercio, A. Cytotoxicity Evaluation of Endodontic Pins on L929 Cell Line. *BioMed Res. Int.* **2019**, *2019*, 5. [CrossRef]
24. Li, W.; Zhou, J.; Xu, Y. Study of the in vitro cytotoxicity testing of medical devices. *Biomed Rep.* **2015**, *3*, 617–620. [CrossRef]
25. Sengottuvelan, A.; Balasubramanian, P.; Will, J.; Boccaccini, A.R. Bioactivation of titanium dioxide scaffolds by ALP-functionalization. *Bioact. Mater.* **2017**, *2*, 108–115. [CrossRef]
26. Zhao, W.; Chang, J. Sol–gel synthesis and in vitro bioactivity of tricalcium silicate powders. *Mater. Lett.* **2004**, *58*, 2350–2353. [CrossRef]
27. Simão, L.; Jiusti, J.; Lóh, N.J.; Hotza, D.; Raupp-Pereira, F.; Labrincha, J.A.; Montedo, O.R.K. Structural Refinement by the Rietveld Method on Clinkers Obtained from Waste from Pulp and Paper Mills. *Mater. Sci. Forum* **2018**, *912*, 175–179. [CrossRef]
28. Menezes, L.R.d.; Silva, E.O.d. The Use of Montmorillonite Clays as Reinforcing Fillers for Dental Adhesives. *Mater. Res.* **2016**, *19*, 236–242. [CrossRef]
29. Lau, D.; Jian, W.; Yu, Z.; Hui, D. Nano-engineering of construction materials using molecular dynamics simulations: Prospects and challenges. *Compos. Part B Eng.* **2018**, *143*, 282–291. [CrossRef]
30. Miguez-Pacheco, V.; Misra, S.K.; Boccaccini, A.R. 4—Biodegradable and bioactive polymer/inorganic phase nanocomposites for bone tissue engineering (BTE). In *Tissue Engineering Using Ceramics and Polymers*, 2nd ed.; Boccaccini, A.R., Ma, P.X., Eds.; Woodhead Publishing: Sawston, Cambridge, UK, 2014; pp. 115–150. [CrossRef]
31. Thevenot, P.; Hu, W.; Tang, L. Surface chemistry influences implant biocompatibility. *Curr. Top. Med. Chem.* **2008**, *8*, 270–280.
32. Anovitz, L.M.; Cole, D.R. Characterization and Analysis of Porosity and Pore Structures. *Rev. Mineral. Geochem.* **2015**, *80*, 61–164. [CrossRef]
33. Tabi, T.; Sajó, I.; Szabó, F.; Luyt, A.; Kovács, J. Crystalline structure of annealed polylactic acid and its relation to processing. *Express Polym. Lett.* **2010**, *4*, 659–668. [CrossRef]
34. Ishii, D.; Ying, T.H.; Mahara, A.; Murakami, S.; Yamaoka, T.; Lee, W.-k.; Iwata, T. In Vivo Tissue Response and Degradation Behavior of PLLA and Stereocomplexed PLA Nanofibers. *Biomacromolecules* **2009**, *10*, 237–242. [CrossRef] [PubMed]
35. Tsuji, H.; Sumida, K. Poly(L-lactide): V. effects of storage in swelling solvents on physical properties and structure of poly(L-lactide). *J. Appl. Polym. Sci.* **2001**, *79*, 1582–1589. [CrossRef]
36. Setiawan, A.H. Determination of Crystallization and Melting Behaviour of Poly-lactic Acid and Polypropyleneblends as a Food Packaging Materials by Differential Scanning Calorimeter. *Procedia Chem.* **2015**, *16*, 489–494. [CrossRef]
37. Wang, M.O.; Etheridge, J.M.; Thompson, J.A.; Vorwald, C.E.; Dean, D.; Fisher, J.P. Evaluation of the in vitro cytotoxicity of cross-linked biomaterials. *Biomacromolecules* **2013**, *14*, 1321–1329. [CrossRef] [PubMed]
38. Knabe, C.; Ducheyne, P. 6—Cellular response to bioactive ceramics. In *Bioceramics and Their Clinical Applications*; Kokubo, T., Ed.; Woodhead Publishing: Sawston, UK, 2008; pp. 133–164. [CrossRef]
39. Nakamura, T.; Takemoto, M. 8—Osteoconduction and its evaluation. In *Bioceramics and their Clinical Applications*; Kokubo, T., Ed.; Woodhead Publishing: Sawston, Cambridge, UK, 2008; pp. 183–198. [CrossRef]
40. Felisbino, S.; Carvalho, H.F. Growth cartilage calcification and formation of bone trabeculae are late and dissociated events in the endochondral ossification of Rana catesbeiana. *Cell Tissue Res.* **2001**, *306*, 319–323. [CrossRef]
41. Sakai, Y.; Kobayashi, M. Lymphocyte 'homing' and chronic inflammation. *Pathol. Int.* **2015**, *65*, 344–354. [CrossRef]
42. Bori, G.; McNally, M.A.; Athanasou, N. Histopathology in Periprosthetic Joint Infection: When Will the Morphomolecular Diagnosis Be a Reality? *BioMed Res. Int.* **2018**, *2018*, 1412701. [CrossRef]

43. Lewis, G. Viscoelastic properties of injectable bone cements for orthopaedic applications: State-of-the-art review. *J. Biomed. Mater. Res. Part B Appl. Biomater.* **2011**, *98*, 171–191. [CrossRef]
44. Vaishya, R.; Chauhan, M.; Vaish, A. Bone cement. *J. Clin. Orthop. Trauma* **2013**, *4*, 157–163. [CrossRef] [PubMed]
45. Omar, O.; Elgali, I.; Dahlin, C.; Thomsen, P. Barrier membranes: More than the barrier effect? *J. Clin. Periodontol.* **2019**, *46*, 103–123. [CrossRef] [PubMed]
46. Turri, A.; Elgali, I.; Vazirisani, F.; Johansson, A.; Emanuelsson, L.; Dahlin, C.; Thomsen, P.; Omar, O. Guided bone regeneration is promoted by the molecular events in the membrane compartment. *Biomaterials* **2016**, *84*, 167–183. [CrossRef] [PubMed]
47. Bai, M.; Yin, H.; Zhao, J.; Li, Y.; Yang, Y.; Wu, Y. Application of PMMA bone cement composited with bone-mineralized collagen in percutaneous kyphoplasty. *Regen. Biomater.* **2017**, *4*, 251–255. [CrossRef] [PubMed]

© 2019 by the authors. Licensee MDPI, Basel, Switzerland. This article is an open access article distributed under the terms and conditions of the Creative Commons Attribution (CC BY) license (http://creativecommons.org/licenses/by/4.0/).

Article

Bioadhesive Matrix Tablets Loaded with Lipophilic Nanoparticles as Vehicles for Drugs for Periodontitis Treatment: Development and Characterization

Denise Murgia [1,2], Giuseppe Angellotti [2], Fabio D'Agostino [3] and Viviana De Caro [2,*]

[1] Dipartimento di Discipline Chirurgiche, Oncologiche e Stomatologiche, Università degli Studi di Palermo, 90127 Palermo, Italy; denise.murgia@unipa.it
[2] Dipartimento di Scienze e Tecnologie Biologiche Chimiche e Farmaceutiche (STEBICEF), Università degli Studi di Palermo, 90123 Palermo, Italy; angellotti.gius@gmail.com
[3] Istituto per lo Studio degli Impatti Antropici e Sostenibilità dell'Ambiente Marino, Consiglio Nazionale delle Ricerche (IAS – CNR), Campobello di Mazara, 91021 Trapani, Italy; fabio.dagostino@cnr.it
* Correspondence: viviana.decaro@unipa.it; Tel.: +39-091-2389-1926

Received: 16 October 2019; Accepted: 30 October 2019; Published: 2 November 2019

Abstract: Periodontitis treatment is usually focused on the reduction or eradication of periodontal pathogens using antibiotics against anaerobic bacteria, such as metronidazole (MTR). Moreover, recently the correlation between periodontal diseases and overexpression of reactive oxygen species (ROS) led to the introduction of antioxidant biomolecules in therapy. In this work, bioadhesive buccal tablets, consisting of a hydrophilic matrix loaded with metronidazole and lipophilic nanoparticles as a vehicle of curcumin, were developed. Curcumin (CUR)-loaded nanostructured lipid carriers (NLC) were prepared using glycyrrhetic acid, hexadecanol, isopropyl palmitate and Tween®80 as a surfactant. As method, homogenization followed by high-frequency sonication was used. After dialysis, CUR-NLC dispersion was evaluated in terms of drug loading (DL, 2.2% w/w) and drug recovery (DR, 88% w/w). NLC, characterized by dynamic light scattering and scanning electron microscopy (SEM), exhibited a spherical shape, an average particle size of 121.6 nm and PDI and PZ values considered optimal for a colloidal nanoparticle dispersion indicating good stability of the system. Subsequently, a hydrophilic sponge was obtained by lyophilization of a gel based on trehalose, Natrosol and PVP-K90, loaded with CUR-NLC and MTR. By compression of the sponge, matrix tablets were obtained and characterized in term of porosity, swelling index, mucoadhesion and drugs release. The ability of the matrix tablets to release CUR and MTR when applied on buccal mucosa and the aptitude of actives to penetrate and/or permeate the tissue were evaluated. The data demonstrate the ability of NLC to promote the penetration of CUR into the lipophilic domains of the mucosal membrane, while MTR can penetrate and permeate the mucosal tissue, where it can perform a loco-regional antibacterial activity. These results strongly support the possibility of using this novel matrix tablet for delivering MTR together with CUR for topical treatment of periodontal diseases.

Keywords: buccal matrix tablets; nanostructured lipid carriers; NLC; metronidazole; curcumin; oral mucosal drug delivery; buccal delivery; hydrophilic sponge; periodontitis; oral disease

1. Introduction

Bacterial infections are among the most common diseases that could affect the oral mucosae. An immunocompromised state of the oral cavity is related to systemic and local factors that could allow the development of infections. Moreover, antibiotic multi-resistance is becoming frequent in bacterial pathogens and it may cause the overgrowth of microorganisms [1].

In particular, periodontal disease is a bacterial infection associated with the overgrowth of anaerobic organisms and, in certain instances with the micro-aerophilic ones, resulting in chronic inflammation of the gingiva [2]. Their action results in progressive connective tissue loss enhanced by the host's inflammatory factors, which can lead to tooth loss [3].

In addition to bacterial action, frequently oral lesions are exasperated by the presence of reactive oxygen species (ROS). Their overproduction that is not balanced by an adequate level of antioxidants can prompt oxidative stress damages. In periodontal disease, most of the damage is induced by inflammatory mediators, involving free radical increasing from the neutrophil/macrophage anti-bacterial response [4]. For this reason, the administration of antioxidant substances in periodontal diseases has recently been extensively studied [5].

Periodontitis treatment is usually focused on the reduction or eradication of periodontal pathogens using antibiotic agents [6].

Metronidazole (MTR) is the first-line antimicrobial agent in the treatment of inflamed periodontal pockets [7], attributable to its activity spectrum against both Gram-negative and Gram-positive anaerobic bacteria and its limited side effects [6,8]. Nonetheless, a prolonged systemic administration of MTR in periodontal disease can raise several issues, such as the increased risk of antibiotic resistance and the presence of side effects like nausea, diarrhea and pseudomembranous colitis [9].

Since oxidative stress is involved as an etiological component of these diseases, a treatment involving MTR together with antioxidant agents could be suitable to obtain synergic action.

Curcumin (CUR) is a natural polyphenolic compound obtained from the rhizome of *Curcuma longa*, where it is present with other characteristic compounds (demethoxycurcumin and bisdemethoxycurcumin) known as curcuminoids [10]. It has been studied for its several pharmaceutical properties, such as being antibacterial [11], antioxidant, anti-inflammatory [12], anti-cancer and a wound healing agent [13].

This compound was frequently used in the treatment of oral lesions, periodontitis in particular, because of its several mechanisms and molecular targets mediating the anti-inflammatory action. It suppresses the activation of the transcription factor NF-κB, cytokines such as TNF-α, IL-1, -2, -6, -8, -12, mitogen-activated protein kinase (MAPK) and c-Jun N terminal kinase (JNK), and downregulates the expression of cyclo-oxygenase-2 (COX-2), lipoxygenase (LOX) and the inducible nitric oxide synthase (iNOS) [14]. Evidences from both in vitro and in vivo studies show that CUR is able to decrease the response of immune cells to periodontal disease-associated bacterial antigens, inhibiting periodontal tissue destruction [15].

CUR is chemically unstable, practically insoluble in water and soluble in solvents such as ethanol and acetone. In the gastrointestinal environment, it suffers slow solubilization and fast degradation that limit absorption. Moreover, a short plasma half-life drastically reduces its action in vivo [16].

Considering the limits of both MTR and CUR systemic administration and the site-specific condition of destructive periodontal disease, a local administration could be an improved treatment for several oral pathologies, since there are minor risks of systemic side effects [17].

Conventional dosage forms may not be accurate due to fast dissolution and disintegration during the application period, while drug delivery systems (DDS) expressly designed to deliver substances in oral mucosa are preferred to promote high concentrations of actives directly into periodontal pockets, improving the in loco treatments. They are designed to stay in contact with the mucosa longer, provide more accurate drug dosing and covering a larger surface area, suited to protect oral wound surfaces [18]. Furthermore, due to its relative immobility, buccal mucosa is a desirable region for retentive systems used for transmucosal drug delivery.

Mucoadhesion is a critical parameter for buccal administration, so materials with optimal adhesive properties, such as mucoadhesive polymers, should be selected. These polymers play an important role in designing bioadhesive oral formulations, since their ability to create strong adhesive contacts with tissues increases the residence time in the target site of such systems [19].

Recently, new mucoadhesive systems, suitable to treat buccal affections, were developed. Buccal tablets are one of the most commonly investigated dosage forms for buccal administration, especially the mucoadhesive ones that give less discomfort in drinking and speaking than the conventional tablet. They can promote drug absorption rates by slow dissolution and erosion [20]. Moreover, molecular size affects the permeability of macromolecules through the oral mucosa, so the drug incorporation as part of nanostructured delivery systems could be developed to enhance the penetration on the target site. Thus, bioadhesive nanoparticles could be more advantageous due to their ability to create intimate contact with larger mucosal surface areas [21].

Nanostructure lipid carriers (NLC) represent a new generation of nanoparticles together with solid lipid nanoparticles (SLN). They are colloidal carrier systems able to deliver active substances, and protect them against chemical degradation, and are mainly used for a controlled release in topical administration [22].

The aim of this work is the formulation, development and characterization of buccal tablets obtained from hydrophilic sponges made by NLC with curcumin (CUR-NLC) entrapped in a dry hydrophilic matrix loaded with MTR. The obtained DDS consists of a bifunctional system able to release CUR and MTR once on the application site, being suitable for the local treatment of several oral diseases.

2. Materials and Methods

2.1. Materials

Curcumin (CUR) was purchased from Alfa Aesar (Haverhill, MA, USA), 1-Hexadecanol (HEXA) from Farmalabor (Canosa di Puglia, Italy) isopropyl palmitate (IP), glycyrrhetic acid (GA) and metronidazole (MTR) from A.C.E.F S.p.a. (Fiorenzuola d'Arda, Italy), Tween 20 and Polyvinylpyrrolidone (PVP-K90) from Sigma-Aldrich (Milan, Italy), Hydroxyethylcellulose (Natrosol™ 250) from Galeno (Comeana PO, Italy) and trehalose dihydrate from la Hayashibara Shoji Inc (Okayama, Japan).

Non-enzymatic artificial plasma of pH 7.4 (PBS) was prepared dissolving 2.80 g of KH_2PO_4 and 20.5 g of Na_2HPO_4 in 1 L of distilled water and the same conditions were used for the non-enzymatic artificial plasma of pH 7.4 (PBS) with 5% and 10% of DMSO.

A buffer pH 6.8 solution simulating salivary fluid was prepared using KCl (1.5 g), $NaHCO_3$ (1.75 g), (KSCN 0.5 g), Na_2HPO_4 H_2O (0.5 g) and lactic acid (1 g) in distilled water; 0.9% saline solution was prepared by dissolving 9 g of NaCl in 1 L of distilled water.

Citrate buffer 10 mM (pH 6.2) used for NLC preparation was obtained dissolving 2.675 g of sodium citrate dehydrate and 0.190 g of citric acid monohydrate in 1 L of distilled water.

All chemicals and solvents of analytical grade were purchased from VWR International (Milan, Italy) and used without further purification. Porcine mucosae were kindly supplied by the Municipal Slaughterhouse of Villabate (Palermo, Italy).

For data processing, we used Kaleidagraph v. 3.5 (Synergy Software Inc, Reading, PA, USA) and Curve Expert 1.34 for Windows as software.

2.2. Lipid Ratio Mixtures Screening Studies

Lipid mixtures (LM) prepared changing both the ratio of the components and CUR are reported in Table n.1.

A calculated amount (1 g) of each LM was melted and stirred using a silicon bath placed on a hot plate (Heidolph MR2001 K equipped with EKT 3001 electronic temperature control, Germany), until the mixture appeared clear and CUR was completely solubilized, then cooled at room temperature and finally stored at 2–4 °C. After 24 h, the melting point of the mixtures was evaluated using Büchi Melting Point B-540 instrument sets at increasing temperature speed of 5 °C/min. The analysis was carried out in triplicate and the result was expressed as the mean of the values obtained.

2.3. Preparation of CUR-Loaded NLC

Aliquots of 100 mg of each lipid mixture were added to 20 mL of 10 mM citrate buffer of pH 6.2 with 0.5% or 1% (w/v) of Tween 20 as a surfactant and homogenized (Polytron Model PT MR

2100 Homogenizer, Kinematica, Luzern, Switzerland) for 1 min. Then a high-frequency sonication treatment using a SONOPULS instrument (mod. HD 2070, Bandelin, Berlin, Germany) at a pulsating frequency of 20 kHz with 0.7 s of activity (on) and 0.3 s of inactivity (off) was performed. Samples were treated with two cycles of sonication (10 min each), the first one at room temperature and the second at 0 °C by immersing the container in an ice/water/NaCl bath.

2.4. NLCs Characterization

2.4.1. Dynamic Light Scattering (DLS) and Z-Potential Measurements

The mean particle size (Z-average), the polydispersity index (PDI) measuring the distribution of the nanoparticle population and the electrical charge on the surface of the nanoparticles (zeta potential, ZP) were determined using a Nano-ZS Zetasizer (Malvern Instruments, Malvern, UK). DLS measures the fluctuations in scattered light intensity due to diffusing particles and represents one of the most powerful techniques to assess particle size [23]. To avoid multiple scatterings of the light caused by a high concentration of particles, samples were prepared by diluting 1 mL of fresh NLC dispersion in 9 mL distilled water. All analyses were performed at 25 °C, under an electrical field of 40 V for ZP analyses, carrying them out in duplicates. Empty NLCs (without CUR) were analyzed to evaluate a potential CUR influence on mean particle size, distribution and stability.

2.4.2. Morphology by Scanning Electron Microscopy (SEM)

SEM analyses were performed to evaluate the morphology and topographic characteristics of both loaded nanoparticles and sponges loaded with MTR and CUR-NLC.

Measurements were carried out using a Zeiss EVO MA10 (Carl Zeiss Microscopy GmbH, Jena, Germany) scanning electron microscope, equipped with a SE-Everhart-Thornley secondary electron detector having as source of electrons a Lanthanum Hexaboride (LaB_6) cathode. The accelerating voltage and probe voltage were, respectively, 20 keV and 10 pA.

The scanning electron microphotographs were acquired in ultra-vacuum condition (HV, about 10^{-7} mbar) and magnified up to 50.000× (200 nm).

To analyze CUR-NLCs, a few drops of fresh CUR-NLC dispersion put on an aluminium stub were placed in a $CaCl_2$ desiccator at 8 °C for 24 h and then coated with an ultrathin layer of gold (thickness about 2 nm) with an AGAR Sputter Coater type system, to increase the surface electrical conductivity. The gold coating was also made for the samples of sponge loaded with MTR and CUR-NLC before each analysis.

2.4.3. Entrapment Efficacy, Drug Loading and Drug Recovery of NLCs

The entrapment efficiency (EE), loading capacity (LC) and Drug Recovery (DR) were indirectly determined, calculating the amount of free CUR in the aqueous phase, applying Equations (1)–(3), respectively [24]:

$$EE\%\left(\frac{w}{w}\right) = \frac{\text{CUR in formulation} - \text{free CUR}}{\text{CUR in formulation}} \times 100, \quad (1)$$

$$DL\%\left(\frac{w}{w}\right) = \frac{\text{CUR in formulation} - \text{free CUR}}{\text{Lipids amount}} \times 100, \quad (2)$$

$$DR\%\left(\frac{w}{w}\right) = \frac{\text{CUR in formulation}}{\text{CUR used}} \times 100. \quad (3)$$

CUR Quantification in Fresh NLC Dispersion

At the end of the NLCs preparation process the volume of the whole formulation was measured. A total of 200 µL of NLC dispersion was withdrawn, transferred into a 10 mL flask and brought to volume with methanol to solubilize all components. Samples were analyzed by HPLC to quantify the total CUR (free and encapsulated) and detect any degradation products [25].

The free CUR, non-entrapped in nanoparticles, was investigated by two methods:

- *Dialysis assay:* Dialysis tube (molecular weight cut off, MWCO, 12–14,000 Da, Visking Dialysis Membrane, Medicell Membranes Ltd., London, UK) was pre-activated and filled by 2 mL of CUR-NLC dispersion and submerged in 350 mL of distilled water, keeping at room temperature and under magnetic stir. After 24 h both the dispersion inside the tube and the external water were analyzed by HPLC.
- *Ultrafiltration assay:* Aliquots of 0.45 mL of fresh samples were centrifuged (Microfuge 22R, Beckman coulter™ Brea, CA, USA) in two Ultrafree-MC (Millipore, Burlington, MA, USA) devices, with membrane cut-off of 10,000 NMWL and 30,000 NMWL, at 8000 rpm and 4 °C for 30 min [26]. In the end, the liquid ultrafiltrate was analyzed by HPLC.

2.5. Preparation of Sponges Loaded with MTR and CUR-NLC.

Firstly, a gel made by trehalose (30%), PVP K90 (0.8%) and Natrosol (5%) was prepared by adding the components to distilled water at 60 °C and keeping in an ultrasonic bath until it appeared dense, clear and completely homogeneous. Afterwards, it was stored at 4 °C for 24 h before use.

Sponges were prepared by adding in two different moments, 100 mg MTR and 4 g of gel to 20 mL of CUR-NLC dispersion. After both the addictions, the sample was gently mixed and then treated with a high-frequency sonication (20 kHz) with cycles of 0.7 s (on) and 0.3 s (off) for 2 min, keeping an ice/water/NaCl bath around it. At the end the sample was stored at −80 °C for a night and then moved in a freeze-dryer (Labconco FreeZone® 2.5 Liter Freeze Dry System) with a 0.014 mPa vacuum for 24 h.

The amount of MTR and CUR loaded in the sponges were determined analyzing three samples of each lyophilized preparation in methanol, by UV-Vis spectrophotometry. Therefore, the morphology of the obtained matrix was evaluated by SEM analysis.

Tablets Preparation

Two steel tablet moulds (PerkinElmer IR Accessory, Waltham, MA, USA) were used to have tablets with different dimensions and weight. Sponges of 20 mg (to obtain Tablet A) and 200 mg (to obtain Tablet B) were placed in molds with a d = 0.47 cm or d = 1.3 cm, respectively. Both were compressed by a 300 g weight for about 30 min. The results are reported as average ± SE ($n = 20$).

2.6. Tablets Characterization

2.6.1. Porosity

Using freeze-drying the solvent is removed, and the space originally occupied by the solvent is retained for the porous structure. The tablet porosity was calculated mathematically, as total pore volume, using the following equation [27]:

$$\%Porosity = \frac{pratical\ volume - theoritical\ volume}{pratical\ volume} \times 100.$$

The bulk volume of the total ingredients of one sponge was calculated according to the equation:

$$Theoritical\ volume = \sum \frac{m_n}{\rho_n}$$

where m and ρ are the mass and density of each ingredient, respectively.

The true volume of tablets was obtained calculating their dimensions as cylinders:

$$Practical\ volume = \pi r^2 \times h.$$

The results of these measurements were also supported by the SEM scanning electron micrographs.

2.6.2. Swelling Test

Swelling tests were performed on B tablets with both weight and optic assessments. For weight tests, tablets were placed on microscope slides and weighted by an analytic balance (Mod. AE 240, Mettler-Toledo S.p.A., Milan, Italy). Every 5 min for 20 min, 0.5 mL of artificial saliva (pH 6.8) were added on the tables and after the removal of the excess water with a filter paper, the weight was assessed. The water uptake was quantified gravimetrically until the weight-plateau or the cleavage of the tablets. Tests were performed on six different tablets and results were reported as means ± SE ($n = 9$; $p < 0.05$).

$$\text{Swelling Index} = \frac{W_0 + (W_t - W_0)}{W_0} \times 100,$$

where W_t is the weight of film at time t, and W_0 is the weight of dry film.

Both plan and frontal optical assessments were performed. Samples placed on glass were positioned on graph paper and wet every 5 min for 2 h by 0.5 mL of simulated saliva (pH 6.8, 37 °C). Each time interval a photograph was taken to evaluate any change in the tablet's morphology.

2.6.3. Ex Vivo Mucoadhesion Strength Measurement

The ex-vivo mucoadhesion strength tests were performed on Tablet B using the modified two armed physical balance method [28]. As model tissue, porcine buccal mucosa excised from slaughtered pigs was used without any pre-treatment. A cyanoacrylate resin (Super Attak Loctite®, Henkel Italia Srl, Milan, Italy) was used to glue a piece of mucosa on a glass support kept in a vessel placed in a thermostatic bath at 37 ± 1 °C. This temperature was maintained during the whole experiment. Before starting the measurements, the mucosa was wetted with 50 µL of simulated salivary fluid. The tablet was fixed using double-sided tape to the lower side of a rubber stopper hanging from the balance arm and then placed just touch the wet mucosal surface; a light force with a fingertip was applied for 20 s.

The measures started 5, 10, 15 and 20 min after tablet placement. The grams required to detach it from the mucosal surface provided the measurement of mucoadhesive strength.

The force of adhesion and the detachment force were calculated using the following equations, respectively [28], and each experiment was performed in triplicate:

$$\text{Force of adhesion } (N) = g \times 9.81/1000,$$

$$\text{Detachment force } \left(N/m^2\right) = \text{Force of adhesion } (N)/\text{surface area } \left(m^2\right).$$

2.6.4. In Vitro CUR and MTR Release Studies

The in vitro CUR and MTR release from tablets was carried out using a Franz diffusion cell (Permeagear, flat flange joint, 9 mm orifice diameter, 15 mL acceptor volume, SES GmbH-Analysesysteme, Bechenheim, Germany). A Cellulose nitrate membrane with a 0.45 µm pore size (Millipore) was soaked with acceptor fluid and fixed between the donor and receptor compartment [29].

The receiver chamber was filled with phosphate buffered to pH 6.2 plus Tween 20 (8% w/v) as a surfactant to increase the solubility and stability of MTR and CUR in the receiver media [30].

The temperature of receptor media was kept at 37 ± 0.5 °C. The donor chamber was filled with Tablet A (20 mg, area = 0.2 cm^2) containing MTR and CUR-NLC and 0.5mL of simulated saliva. At regular time intervals (15 min), samples (0.5 mL) were withdrawn from the acceptor compartment and the sample volume was taken out and replaced with fresh fluid. The samples were analysed by UV-Vis using the appropriate blank and calibration curve. The permeation experiments were carried out for 3 h and results were reported as means ± SE ($n = 6$). Release data were elaborated using Kaleidagraph v. 3.5 software and fitted to the semi-empirical equation that is usually applied to describe drug release from polymeric systems [31]. Linear or non-linear least squares fitting methods were used to determine the optimum values for the parameters present in each equation. Fittings were validated by using χ^2. A p-value of less than 0.05 was considered to be statistically significant.

2.7. Ex Vivo Permeation and Penetration of CUR and MTR throughout Porcine Buccal Mucosa

The permeation or/and penetration of CUR and MTR released from tablets through the porcine buccal mucosa, was evaluated using Franz type diffusion cells (Permeagear, flat flange joint, 9 mm orifice diameter, 15 mL acceptor volume, SES GmbH-Analysesysteme, Bechenheim, Germany), used as a two-compartment open model. Mucosal specimens (kindly supplying by Municipal Slaughterhouse of Villabate, Palermo, Italy) consisted of tissue removed from the vestibular area of the retromolar trigone (buccal mucosa) of freshly slaughtered domestic 6–8-month-old pigs were used as a membrane. They were prepared as described previously [32].

The specimens, transferred in the laboratory within 2 h from animal sacrifices, were surgically treated to remove the excess of adipose and connective tissues and then stored at −20 °C for periods up to one week. Before the experiments they were equilibrated at room temperature and dipped for about 1 min in saline solution at 60 °C; the connective tissue was carefully peeled off from the mucosa (slides 250 ± 25 µm and 100 ± 25 µm thick for buccal and sublingual tissues, respectively) to obtain the heat-separated epithelium along with the intact basal lamina [33].

The thickness was measured using a digital micrometer. After this heath treatment, which is able to not modify the permeability and/or integrity of the mucosae [32], specimens were equilibrated in PBS for about 3 h at room temperature to remove biological matter that could interfere with analysis. The equilibration medium was replaced with fresh PBS every 15 min.

The Franz diffusion cells were equilibrated for 30 min at 37 ± 0.5 °C, mounted with buccal mucosa as membrane, PBS (pH 7.4) with 10% DMSO as acceptor compartment and 1 mL and simulated saliva in donor chamber. This step was followed by the removal of donor compartment solution, immediately replaced with a Tablet A (area = 0.2 cm^2) containing MTR and CUR-NLC and 0.4mL of simulated saliva. At regular time intervals (30 min), samples (0.5 mL) were withdrawn from the acceptor compartment and the sample volume was taken out and replaced with fresh fluid. The permeation experiments were carried out for 4 h. The CUR and MTR amount in acceptor chamber were quantitatively determined by UV-Vis spectrophotometry using the appropriate blank and calibration curve and results were reported as means ± SE ($n = 12$).

At the end of each experiment, mucosal integrity was checked as previously described [34]. The flux values (Js) across the membranes were calculated at the steady state per unit area by linear regression analysis of permeation data, following the relationship $Js = Q/At$ (mg/cm^2·h), where Q is the amount of drugs recovered in the acceptor compartment, A is the cross-sectional area available for diffusion (0.636 cm^2), and t is the time of exposure (h). Data were elaborated using Kaleidagraph 3.5 (Sinergy Software Inc., Reading, PA, USA) as software.

Quantification of CUR and MTR Entrapped into Porcine Mucosa

At the end of each experiment, the residual MTR and CUR amount entrapped into the mucosal tissue were quantified by extraction. Each mucosa specimen was washed with PBS (3 × 2 mL) and was then dipped for 5 min in warmed (50 °C) methanol (1.5 mL). The extraction was repeated three times and the collected mother liquors were quantitatively transferred in a 5 mL flask and brought to volume. The amount of drugs extracted was evaluated by UV and HPLC analysis using the appropriate calibration curve and blank. The same extraction treatment was also performed on mucosal specimens subjected to an experimental phase in the absence of drugs and was used as the control. Data were reported as percentage amount of CUR and MTR into the tissue. Moreover, the Accumulation (Ac) parameter was calculated using the following equation:

$$Ac\ (cm) = [Q\ (mg)/A\ (cm^2)]/Cd\ (mg/cm^3)$$

where Q is the amount of drug which remains entrapped into the buccal tissue at the end of permeation experiments, A is the active cross-sectional area available for diffusion (0.636 cm^2), and Cd is drug concentration in the donor compartment (mg/cm^3) [35].

2.8. Drugs Assay

2.8.1. UV-Vis Method

The amount of drugs loaded in NLCs and sponges, as well as drugs detected during release and permeation experiments, were measured spectrophotometrically using the appropriate calibration curves and blanks (UV/VIS mod. Pharma Spec 1700, Shimadzu, Tokyo, Japan) with simple, accurate and reproducible methods.

For CUR determination, for λ_{max} = 428 nm, two calibration curve in methanol were performed: In the linearity range of 0.0002–0.0075 mg/mL, the regression equation was Abs = −0.0289 + 132x (mg/mL) (R = 0.999, SE 0.0101) and in the linearity range of 0.0025–0.01 mg/mL, the regression equation was Abs = −0.0108 + 136 x (mg/mL) (R = 0.997, SE 0.0397).

The calibration curve in the phosphate buffer of pH 6.2 plus Tween 20 (8% w/v) was in the linearity range of 0.0001–0.001 mg/mL, and the regression equation was Abs = 0.01277 + 160x (mg/mL) (R = 0.997, SE 0.0105).

For MTR determination were found validation parameters in methanol, in PBS and in phosphate buffer (pH 6.2) plus Tween 20 (8% w/v), at λ_{max} = 310 nm. For methanol, the linearity range was 0.0025–0.03 mg/mL, and the regression equation Abs = 0.0011+ 62x (mg/mL) (R = 0.999, SE 0.0092); for PBS the linearity range was 0.0025–0.0125 mg/mL, and the regression equation Abs = 0.0011 + 61.78x (mg/mL) (R = 0.999, SE 0.0040). For the phosphate buffer (pH 6.2) plus Tween 20 (8% w/v), the linearity range was 0.0025–0.02 mg/mL, and the regression equation was Abs = 0.0771 + 46.8x (mg/mL) (R = 0.998, SE 0.0102).

At the testing concentrations, no interferences between drugs and components of formulations were observed and no change in drug absorbance at its λmax was experienced in the presence of excipients. In analogy, the amount of drugs founded in acceptor and/or entrapped into the membrane were measured after withdrawal or extraction from mucosal tissue by methanol. Intraday and interday variations, observed during the collection of experimental data, were lower than sensibility.

2.8.2. HPLC Method

HPLC analyses were performed with an HPLC Shimadzu LC-10AD VP instrument (Tokyo, Japan) equipped with a binary pump LC-10AD VP, a UV SPD-M20A Diode Array detector, a 20 µL injector and a computer integrating apparatus (EZ Start 7.4 software, Shimadzu Scientific Instruments, Inc., Columbia, MD, USA). Chromatographic separation was achieved on a reversed-phase column ACE® EXCEL 5 C18-AMIDE (5 µm, 4.6 mm × 125 mm), a mobile phase consisted of methanol (A) acetate buffer 5 mM pH 6.5 (B). For separation the gradient method was developed as follows: A: B (0.5:99.5→0.01–3.00 min, 80:20→3.00–7.00 min; 80:20→7.00–22.00 min; 0.5:99.5→22–26 min). The flow rate was set at 1 mL/min, the UV wavelength range 200–700 nm and set 428, 310 and 250 nm to identification of MTR, CUR and GA, respectively. In these conditions the retention time of MTR, CUR and GA were 1.95, 12.42 and 15.68 min, respectively.

For CUR quantification, the calibration curve was performed in the concentration range of 0.0075–0.01 mg/mL. HPLC reports were highly reproducible and linearly related to concentration (R = 0.999).

2.9. Data Analysis

Data were expressed as mean ± SE. For the statistical analysis of the data, Student's *t*-test has been applied with the minimum levels of significance with $p < 0.05$.

3. Results and Discussion

3.1. CUR-NLC Formulation and Characterization

In the pharmaceutical field, nano-encapsulation is highly exploited to promote the administration of substances with a low ability to cross biological membranes, using natural and/or synthetic materials to have a formulation with suitable chemical–physical characteristics. The low stability and solubility in a water solution of curcumin involve its delivery by NLCs [25].

Therefore, as components for lipid mixtures were chosen, long-chain fatty alcohol, such as 1-Hexadecanol, isopropyl palmitate, a liquid excipient able to solubilize lipophilic drugs, and 18-β glycyrrhetic acid (GA) with emulsifying properties. GA is a triterpenoid metabolite of glycyrrhizin, extracted from the dried roots and rhizomes of *Glycyrrhiza glabra* [36], studied for its anti-ulcer, antihepatotoxic, anti-microbial, anti-fungal, antitussive, anti-diabetic, anti-diuretic, skin whitening, cytoprotective and cytotoxic activities [37]. Moreover, it is well known as antioxidant tested against b-carotene destruction and LDL oxidation [38] and as anti-inflammatory agent, used for treating acute and chronic dermatitis reducing skin lesion sizes [39].

In order to prepare CUR-NLCs with high loading capacity and good stability is crucial the choose of a lipid mixture able to permit CUR solubilization and to confer to the whole mixture (lipids and CUR) a melting point similar to the physiological temperature of the oral cavity. Indeed, once applied on the mucosal target sites, NLC should undergo partitioning into the tissue and promote drug penetration.

Lipid screening study results (Table 1) showed how the ratios of high-melting and low-melting components affect the final melting point of the mixes.

Table 1. Lipid mixture compositions.

Sample	CUR %	GA %	HEXA %	IP %	Melting Point °C
Mixture 1	1	1	39.2	58.8	31 ± 2 °C
Mixture 2	1	1	58.8	39.2	40 ± 2 °C
Mixture 3	1	3	58	38	45 ± 2 °C
Mixture 4	2.5	3	56.7	37.8	48 ± 2 °C
Empty	0	3	57.3	38.2	42 ± 2 °C

Mixtures 1 and 2 had to be discarded for the low melting point, due to the high amount of IP, while Mixtures 3 and 4 presented melting points suitable for NLCs preparation.

In particular, CUR up to 2.5% w/w showed completely solubilized in Mixture 4, which appeared clear in the molten state.

Regarding the NLC preparation method, a homogenization followed by high-frequency sonication was performed. A hot pre-emulsion of melted lipid mixture and the aqueous emulsifier phase is obtained by hot homogenization, carried out at temperatures above the melting point of the lipids. A citrate buffer at pH 6.2 was chosen as an aqueous medium due to the pH-dependent instability of CUR showed as decomposition at a neutral or basic pH [40]. To reduce the particle size, sonication by a high-intensity ultrasonic probe was performed at the same temperature. Subsequently, the obtained hot nano-emulsion was treated by the second cycle of ultrasonication at 0 °C, which leads to the crystallization of the lipids and NLC formation. All the obtained dispersions appeared stable, homogeneous and easily to re-disperse.

Particle size, size distribution and Zeta potential are essential factors to value the stability of colloidal systems, showing significant effects on the final nanoparticles' behavior [41].

Table 2 shows the DLS measurements of average particle size, particle size distribution and electrical charge surface of the nanoparticles expressed respectively as Z-average, PDI and Z-potential.

Table 2. Results of dynamic light scattering (DLS) and Z-Potential measurements.

.	Tween 20%	Z-Average (nm)	PDI	Z-Potential (mV)
Mixture 3	0.5	100	0.242	−24.6
	1	100	0.274	−24.4
Mixture 4	0.5	121.6	0.235	−37.4
	1	121.8	0.272	−33.2
Empty	0.5	129.4	0.216	−8.4

The surfactant amount is an important factor to determine the nanoparticle physicochemical characteristics due to its surface-active properties. To prepare NLCs with a smaller particle size, the concentration of surfactant could be increased in order to decrease the interfacial tension between lipid and aqueous phases. From Mixtures 3 and 4 were prepared CUR-NLCs with 0.5% or 1% of Tween 20, to evaluate any difference whether in particle size or system stability. The values were comparable and they clearly suggested that the lowest percentage of surfactant was sufficient to cover the surface of nanoparticles and prevent the lipid agglomeration during the homogenization process [42]. In particular, this stability may be enhanced by both the electrostatic repulsion between particles with the same charge and the stabilization and steric hindrance of Tween 20 chains. Although NLC containing 1% CUR showed smaller particle size and a good PDI value, the NLC from Mixture 4, showing suitable average particle size, the PDI value that indicates a narrow size distribution and Z-potential that reflects great physical stability of the dispersion, was preferred due to the higher CUR load. These results demonstrate that the amount of CUR loaded into the NLC plays a fundamental role in particle size.

The empty samples were also analyzed to understand the CUR contribution in system stability. The lack of CUR caused an increase in the ZP value, while the other parameters did not change significantly. However, this value suggests a low physical stability.

According to the last considerations, the formulations made by Mixture 4 and 0.5% Tween 20 were chosen for the next characterizations.

To obtain information concerning the surface characteristics and the morphology of CUR-NLC, SEM analysis was also performed and scanning electron micrographs are presented in Figure 1. NLCs show uniform spherical shapes with smooth surfaces, and their sizes was found evenly within the nanometer range. It is observable that the agglomeration of particles may be due to the lipid nature of carriers and the sample treatment prior the SEM analysis.

The drug recovery (DR) of CUR was calculated from the effective volume of fresh NLCs dispersion obtained (about 18 mL) with respect to the volume of buffer initially used (20 mL), where CUR resulted in 0.11 mg/mL, determining the 88% of DR. Furthermore, no degradation products were detected during CUR quantification by HPLC analyses.

EE and DL were indirectly determined, evaluating the amount of free CUR remaining left outside the nanoparticles by the dialysis and the ultrafiltration assays. In both the analyses no free CUR was found, so the EE was 100%. The DL calculated was of 2.2% (initial CUR 2.5%) that may be due to the loss of volume of the sample during the preparation process.

However, the high EE and DL values confirm the solubility of CUR in the chosen lipids as well as the suitable formulation design for CUR encapsulation.

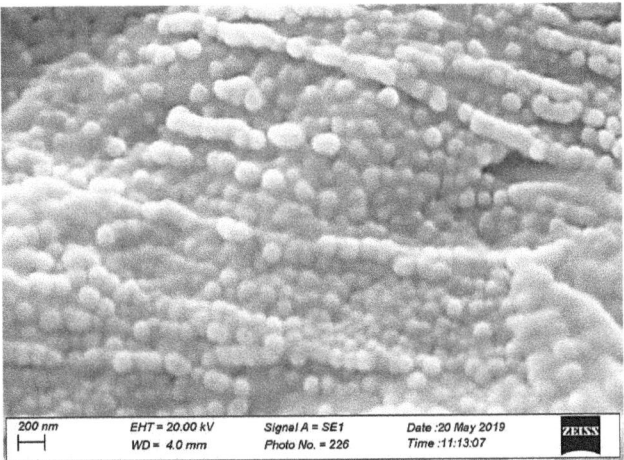

Figure 1. SEM morphology of CUR–NLC, showing the surface structure of the dried dispersion of CUR–NLC. Bar = 200 nm.

3.2. Sponges Loaded with MTR and CUR-NLCs and Tablet Formulation and Characterization

Sponges were prepared entrapping the CUR-NLCs and MTR in a mucoadhesive hydrophilic polymeric matrix subsequently dried.

Natrosol^{TM}, a nonionic, biocompatible, mucoadhesive, nontoxic polymer was chosen to prepare a hydrogel together with PVP-90 as mucoadhesive plasticizer and trehalose as cryoprotectant. These excipients were selected for the suitable lyophilization behavior and the acceptable physical appearance of the resulting products.

The freeze-dried samples appeared soft, porous and friable. Besides the matrix morphology, the inner structure and surface topography were evaluated by SEM (Figure 2). The images showed an irregular structure with varying pore sizes and a high porosity that allows a more rapid entrance of water during dissolution that might mean an increased drug release rate. No drug crystal or aggregation of particles was visible in the photograph.

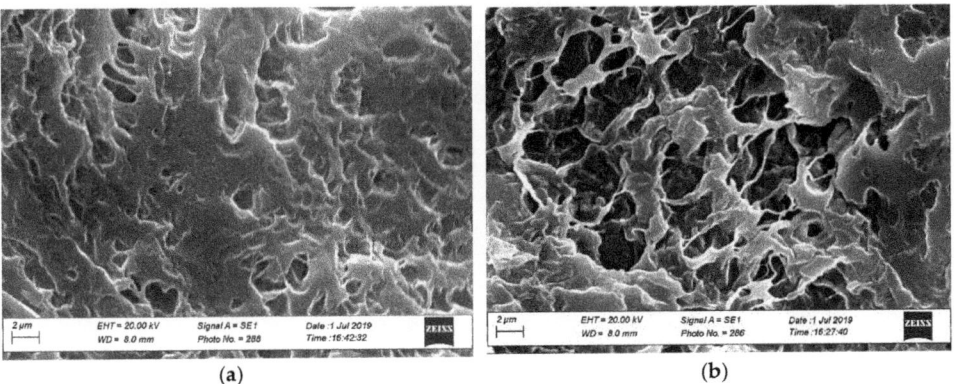

Figure 2. SEM of surface morphology and internal structure of a hydrophilic matrix (**a**) and hydrophilic matrix loaded with CUR-NLC and MTR (**b**). Bar = 2 μm.

The homogeneous distribution of MTR and CUR into the lyophilized matrix was verified withdrawing portions of powder of the same batch and quantifying the drugs by UV-Vis analysis.

The results showed an average content of MTR and CUR of 9.57% and 0.25% (w/w), respectively.

The tablets have been designed to adhere and swell once applied on the mucosae, promoting an adequate penetration of both hydrophilic and hydrophobic actives. Indeed, they allow a precise location and adhesion on the target site, an extended retention time of formulation and an enhanced delivery in situ of actives.

Buccal tablets were prepared from the compression of aliquots of sponges obtained after the freeze-drying process. Tablets of 20 mg (A) and 200 mg (B) of sponge were made to perform different characterization experiments.

The reproducibility of the tablet's preparation has been assessed measuring the average weight and thickness of both A and B ones. They were respectively 20 ± 0.3 mg, 2.34 ± 0.05 mm for Tablet A and 200 ± 1.2 mg 3.31 ± 0.05mm for Tablet B. All data are following the requirements of the Italian Pharmacopoeia (F.U. XII ed.) and confirm high product reproducibility.

After compression, they preserved the porous structure generated during the freeze-drying process, with the solvent removal. When the structure surrounding the liquid is rigid enough to prevent pore collapse, the pore structure is retained. So, the remaining space originally occupied by the solvent becomes pores in the scaffold. By investigation, the porosity of the tablets was calculated to be 96, 8%. This value was perfectly in accordance with the spongy matrix shown in the images obtained by SEM analyses.

For a suitable behavior and a sustained and controlled drug release, water penetration and tablet hydration capacity are critical properties here evaluated performing swelling tests as weight and optic assessments. Weight variations of the tablets are expressed as perceptual weight increase versus time, and they are reported in Figure 3. The data showed a rapid increase in weight in the first 20 min-interval, equal to 60% of the initial one, due to the absorption of saliva and the consequent swelling. After this time a much slower fluid absorption follows due to the formation of a viscous gel layer (Figure 4) that remains on the mucosa until a slow erosion phenomenon occurs.

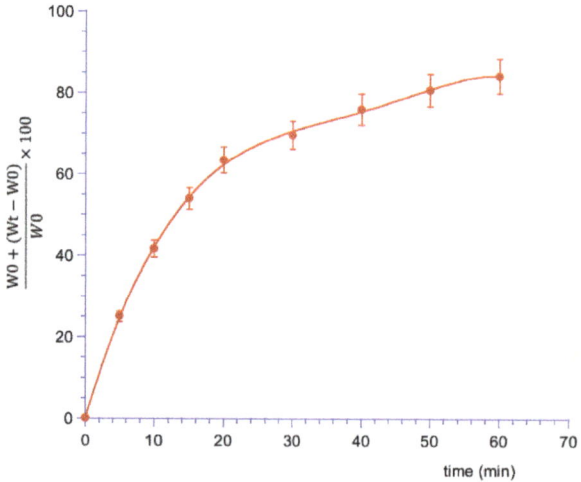

Figure 3. Swelling index measured as percent of weight increased vs. time. Values are presented as means ± SE ($n = 9$).

Figure 4. Tablet after 60 min of contact with simulated saliva.

The plan and frontal optical assessments (Figure 5) were carried out to measure the surface area of the oral mucosa, which is effectively covered by the tablet when dissolution occurs once it was applied on the target site. Both the frontal and plan photographs showed that the swelling of the tablet in the first 20 min was due to the main absorption of saliva that does not involve a large increase in volume and a partial loss of weight, and dissolution occur in 120 min.

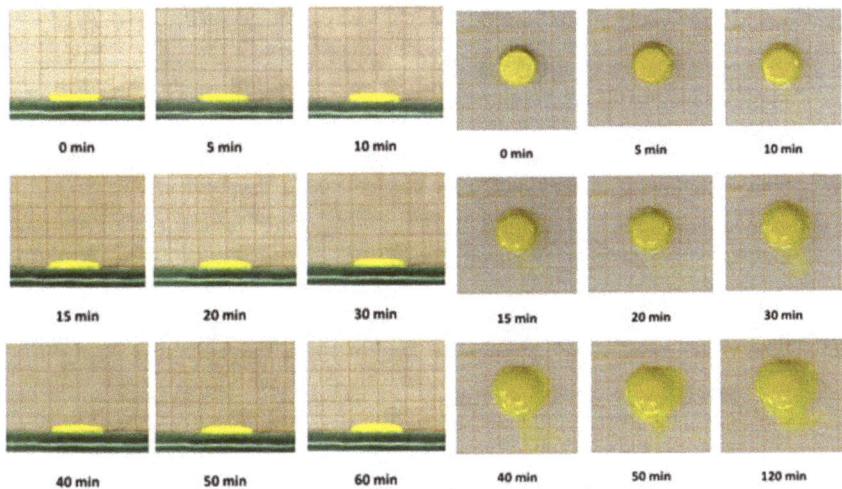

Figure 5. Timeline from 0 to 120 min of the plan and frontal swelling of the tablet.

3.3. Ex Vivo Mucoadhesion Strength Measurement

The mucoadhesive properties of buccal tablets influence the ability of the dosage form to be retained at the site of action, in intimate contact with the target site. For designing mucoadhesive tablets, polymers as polyacrylic acid and cellulose derivatives are often chosen for their suitable physical and chemical properties. Ex vivo mucoadhesion strength measurements were performed in triplicate by a modified two-armed physical balance, applying the tablets on fresh-cut porcine buccal mucosa. Variation in adhesive strength as a function of contact time was evaluated, and as is shown in Table 3, the mucoadhesive force increased with the increase of the contact time of formulation with mucin.

These suitable mucoadhesive properties are enhanced by the use of Natrosol and PVP-K90, polymers able to form addition bonds with mucins increasing the residence time of formulation on target mucosal site.

Table 3. Force of adhesion and detachment force of the tablet after different contact time on the porcine buccal mucosa ($n = 3$).

Contact Time (min)	Force of Adhesion (N)	Detachment Force (N/m^2)
5	0.064	486.81
10	0.073	549.50
15	0.100	752.35
20	0.146	1095.33

The release of drug molecules or nanoparticles from prepared tablets is necessary for their transport into and across the buccal membrane to exhibit therapeutic activity. However, the release of drugs from a swellable matrix depends on water diffusion into the matrix, polymer swelling and gel formation, then drug diffusion through the gel or its liberation from the gel [43].

Figure 6 analyses the dose fraction of CUR and MTR released from Tablet A at specific time points. The period of the drug release testing was kept for 3 h, as about 60% of MTR and 17% of CUR release were observed during this time.

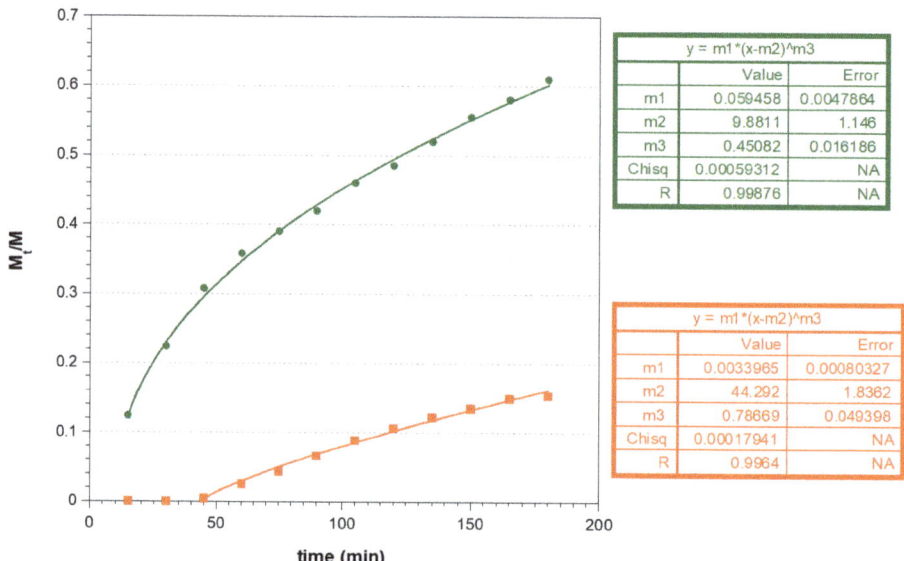

Figure 6. Plot of cumulative amount of MTR (●) and CUR (■) released from Tablet A and their fit with the modified Korsmeyer–Peppas model of MTR (–) and CUR (–).

Different physico-chemical phenomena are involved in the matrix tablet release mechanism. From the data, it is possible to highlight that there is a lag time in the release of CUR, probably because the matrix must initially absorb water and subsequently swell, then releasing the NLC containing CUR. This phenomenon is much less evident in the release profile of the MTR as it dissolves immediately in the fluid absorbed by the tablet and diffuses through it. By the fitting the mathematical models that are useful for comprehension of all the phenomena affecting drug release kinetics, the modified form of Korsmeyer–Peppas model, as described:

$$\frac{M_t}{M_\infty} = K(t-l)^n$$

where M_t is the amount of drug released over time t, M_∞ is the amount of drug contained in the dosage form at the beginning of the release process, K is the constant of incorporation of characteristics of the system, n is the exponent of release (related to the drug release mechanism and geometry of dosage form) in function of time t and l is the latency time, which marks the beginning of drug release from the system; the most appropriate to describe the behavior of our matrix tablet (Figure 6).

Experimental n values obtained by fitting can explain the complex phenomena that occur during MTR and CUR-NLC release from the tablet. Possibly, the MTR behavior follows a Fickian model (n = 0.45, for cylinders) and the release is governed primarily by diffusion. MTR is readily solubilized, and the solvent transport rate or diffusion is much greater than the process of polymeric chains relaxation. For CUR-NLC release, the value n = 0.77 indicates that the model is non-Fickian or anomalous transport, and the mechanism of drug release is governed by both diffusion and swelling. This could be attributed to the hydrophobicity or dimensions of the nanoparticles, so that their diffusion through the matrix is influenced by the swelling or relaxation of polymeric chains.

3.4. Ex Vivo Permeation and Penetration of CUR and MTR throughout Porcine Mucosa

The ability of the tablets to release CUR and MTR and the aptitude of actives to penetrate and/or permeate the membrane was evaluated using vertical Franz type diffusion cells and porcine buccal mucosa, a useful model to simulate human epithelium [34].

The spectrophotometric analyses demonstrate that just MTR permeates the buccal membrane, indicating its increasing concentration in the acceptor compartment. The drug movements from tablets to artificial plasma expressed as an accumulative amount of permeated MTR versus time, are showed in Figure 7.

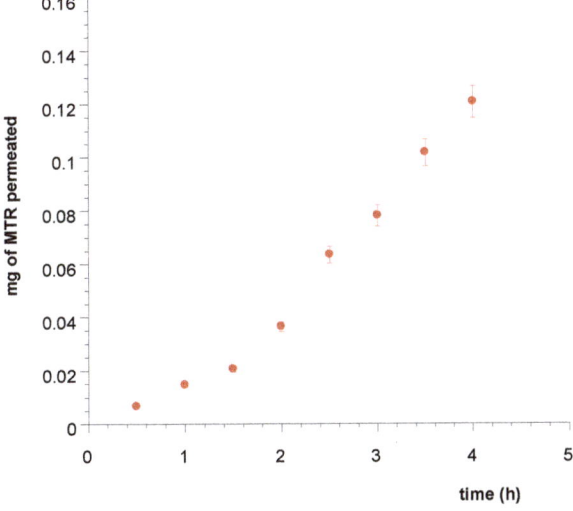

Figure 7. Plot of the cumulative amount of MTR permeated across porcine buccal mucosa vs. time from Tablet A soaked with simulated saliva. Values are presented as means ± SE (n = 12).

Tablets applied on porcine buccal mucosa for four hours produced massive input of MTR in the acceptor compartment and the extrapolated flux (Js) per unit area of MTR through the mucosal membrane at the steady state resulted 0.0731 mg/cm^2 h (Figure 8). However, the experiments of drug permeation occurred from just one side of the tablet adhered on the mucosa, while in vivo conditions may imply that each side of tablets is in contact with the mucosa. So, a tablet of 1 cm^2 produces a drug flux of about 0.146 mg/cm^{2}h.

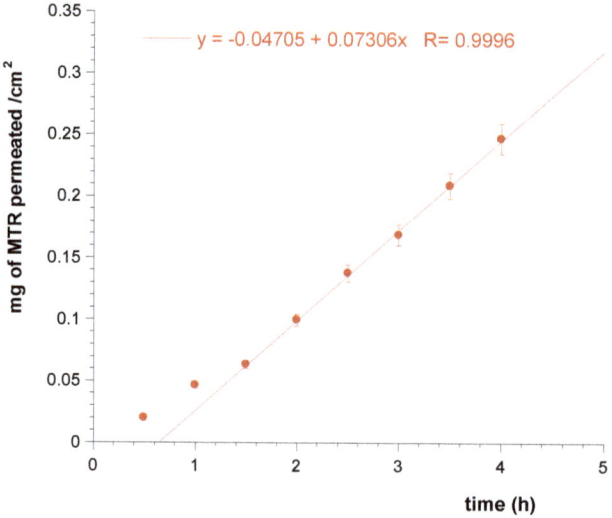

Figure 8. Linear fit at the steady state of MTR permeation per cm^2 of porcine sublingual mucosa.

The residual CUR and MTR entrapped into the membrane were quantified by methanol extraction and analyzed respectively by HPLC and UV-Vis assays. After 4 h, the average amounts extracted were 0.0032 mg of CUR and e 0.0835 mg of MTR, that are respectively 6.31% and the 4.37% of the dose applied. These results demonstrate the ability of NLC to promote the penetration of CUR through the lipophilic domains of the mucosal membrane, where its accumulation can carry out its antioxidant activity. Moreover, results show that MTR is also able to penetrate and permeate the mucosal tissue, where it can perform a loco-regional antibacterial activity. The Ac parameter calculated for both the actives, 0.1006 and 0.0687 cm for CUR and MTR, respectively, demonstrated that the lipophilic molecule CUR possesses great tendency to penetrate lipophilic domains of membrane and accumulate into that. On the other hands, CUR is very scarcely soluble in aqueous fluids so that, without lipid nanocarriers, it cannot penetrate the tissue. By contrast, the MTR having a fair solubility in aqueous fluids, and due to its partition coefficient, can diffuse through the tissues and be distributed both in aqueous and lipophilic domains.

4. Conclusions

In this work, a novel buccal drug delivery system for the topical treatment of oral diseases related to both oxidative stress and bacterial overgrowth was designed. Nanostructured lipid carriers containing CUR have been formulated, characterized and successfully incorporated into a mucoadhesive hydrophilic sponge also containing MTR. The buccal tablets obtained by soft compression of sponge showed an appropriate porosity and swelling index as well as mucoadhesion strength. Ex vivo studies have shown that the tablets are able to release the actives promoting CUR penetration and MTR permeation. The obtained results encourage further studies about the possibility to use these matrix systems to deliver in situ antioxidant and antimicrobial agents to treat several diseases that affect oral mucosae. Finally, this matrix tablet was specifically designed to contrast oral infections, for which MTR is the drug of choice. However, this new DDS could be loaded with other drugs and potentially be useful for various pathologies affecting other mucous membranes (e.g., vaginal, rectal or nasal).

Author Contributions: V.D.C. conceived and designed the project, interpreted the results and edited the manuscript; D.M. and G.A. performed the experiments, analyzed the data and prepared the initial draft of the manuscript; F.D. performed morphology studies by SEM and analyzed the data.

Funding: This research was funded by "Funding of Research Base 2017" (FFABR-2017), MIUR, Italy.

Acknowledgments: Gaspare Orlando, from the Municipal Slaughterhouse of Villabate (Palermo, Italy) for kindly supplied Porcine mucosae.

Conflicts of Interest: The authors declare no conflict of interest.

References

1. Dahlén, G. Bacterial infections of the oral mucosa. *Periodontol. 2000* **2009**, *49*, 13–38. [CrossRef]
2. Hoare, A.; Soto, C.; Rojas-Celis, V.; Bravo, D. Chronic Inflammation as a Link between Periodontitis and Carcinogenesis. *Mediat. Inflamm.* **2019**, *2019*, 1029857. [CrossRef] [PubMed]
3. Cekici, A.; Kantarci, A.; Hasturk, H.; Van Dyke, T.E. Inflammatory and immune pathways in the pathogenesis of periodontal disease. *Periodontol. 2000* **2014**, *64*, 57–80. [CrossRef] [PubMed]
4. Kesarwala, A.H.; Krishna, M.C.; Mitchell, J.B. Oxidative stress in oral diseases. *Oral Dis.* **2016**, *22*, 9–18. [CrossRef] [PubMed]
5. Murgia, D.; Mauceri, R.; Campisi, G.; De Caro, V. Advance on resveratrol application in bone regeneration: Progress and perspectives for use in oral and maxillofacial surgery. *Biomolecules* **2019**, *9*, 94. [CrossRef]
6. Sato, S.; Fonseca, M.J.V.; Del Ciampo, J.O.; Jabor, J.R.; Pedrazzi, V. Metronidazole-containing gel for the treatment of periodontitis: An in vivo evaluation. *Braz. Oral Res.* **2008**, *22*, 145–150. [CrossRef]
7. Labib, G.S.; Aldawsari, H.M.; Badr-eldin, S.M. Metronidazole and Pentoxifylline films for the local treatment of chronic periodontal pockets: Preparation, in vitro evaluation and clinical assessment. *Expert Opin. Drug Deliv.* **2014**, *11*, 855–865. [CrossRef]
8. Lo, S.; Edlund, C.; Nord, C.E. Metronidazole Is Still the Drug of Choice for Treatment of Anaerobic Infections. *Clin. Infect. Dis.* **2010**, *50*. [CrossRef]
9. Brook, I.; Wexler, H.M.; Goldstein, E.J.C. Antianaerobic antimicrobials: Spectrum and susceptibility testing. *Clin. Microbiol. Rev.* **2013**, *26*, 526–546. [CrossRef]
10. Zheng, Q.T.; Yang, Z.H.; Yu, L.Y.; Ren, Y.Y.; Huang, Q.X.; Liu, Q.; Ma, X.Y.; Chen, Z.K.; Wang, Z.B.; Zheng, X. Synthesis and antioxidant activity of curcumin analogs. *J. Asian Nat. Prod. Res.* **2017**, *19*, 489–503. [CrossRef]
11. Gupta, S.C.; Patchva, S.; Aggarwal, B.B. Therapeutic roles of curcumin: Lessons learned from clinical trials. *AAPS J.* **2012**, *15*, 195–218. [CrossRef] [PubMed]
12. Menon, V.P.; Sudheer, A.R. *Antioxidant and Anti-Inflammatory Properties of Curcumin*; Aggarwal, B.B., Surh, Y.-J., Shishodia, S., Eds.; Springer: Boston, MA, USA, 2007; Volume 595.
13. Akbik, D.; Ghadiri, M.; Chrzanowski, W.; Rohanizadeh, R. Curcumin as a wound healing agent. *Life Sci.* **2014**, *116*, 1–7. [CrossRef] [PubMed]
14. Jacob, P.; Nath, S.; Sultan, O. Use of Curcumin in Periodontal Inflammation. *Interdiscip. J. Microinflamm.* **2014**, *1*. [CrossRef]
15. Dave, D.H.; Patel, P.; Shah, M.; Dadawala, S.M.; Saraiya, K.; Sant, A.V. Comparative Evaluation of Efficacy of Oral Curcumin Gel as an Adjunct to Scaling and Root Planing in the Treatment of Chronic Periodontitis. *Adv. Hum. Biol.* **2018**, 64–69. [CrossRef]
16. Jain, S.; Rama, S.; Meka, K.; Chatterjee, K. Curcumin eluting nanofibers augment osteogenesis toward phytochemical based bone tissue engineering. *Biomed. Mater.* **2016**, *11*, 055007. [CrossRef]
17. Joshi, D.; Garg, T.; Goyal, A.K.; Rath, G. Advanced drug delivery approaches against periodontitis. *Drug Deliv.* **2016**, *23*, 363–377. [CrossRef]
18. De Caro, V.; Scaturro, A.L.; Di Prima, G.; Avellone, G.; Sutera, F.M.; Di Fede, O.; Campisi, G.; Giannola, L.I. Aloin delivery on buccal mucosa: Ex vivo studies and design of a new locoregional dosing system. *Drug Dev. Ind. Pharm.* **2015**, *41*, 1541–1547. [CrossRef]
19. De Caro, V.; Ajovalasit, A.; Sutera, F.M.; Murgia, D.; Sabatino, M.A.; Dispenza, C. Development and Characterization of an Amorphous Solid Dispersion of Furosemide in the Form of a Sublingual Bioadhesive Film to Enhance Bioavailability. *Pharmaceutics* **2017**, *9*, 22. [CrossRef]
20. Laffleur, F. Mucoadhesive polymers for buccal drug delivery. *Drug Dev. Ind. Pharm.* **2014**, *40*, 591–598. [CrossRef]
21. Morantes, S.J.; Buitrago, D.M.; Ibla, J.F.; García, Y.M.; Lafaurie, G.I.; Parraga, J.E. 5- Composites of hydrogels and nanoparticles: A potential solution to current challenges in buccal drug delivery. In *Biopolymer-Based Composites*; Jana, S., Maiti, S., Jana, S., Eds.; Elsevier Ltd.: Amsterdam, The Netherlands, 2017; pp. 107–138.

22. Müller, R.H.; Radtke, M.; Wissing, S. A Solid lipid nanoparticles (SLN) and nanostructured lipid carriers (NLC) in cosmetic and dermatological preparations. *Adv. Drug Deliv. Rev.* **2002**, *54* (Suppl. 1), S131–S155.
23. Marques, A.C.; Rocha, A.I.; Leal, P.; Estanqueiro, M.; Lobo, J.M.S. Development and characterization of mucoadhesive buccal gels containing lipid nanoparticles of ibuprofen. *Int. J. Pharm.* **2017**, *533*, 455–462. [CrossRef] [PubMed]
24. Hazzah, H.A.; Farid, R.M.; Nasra, M.M.A.; EL-Massik, M.A.; Abdallah, O.Y. Lyophilized sponges loaded with curcumin solid lipid nanoparticles for buccal delivery: Development and characterization. *Int. J. Pharm.* **2015**, *492*, 248–257. [CrossRef] [PubMed]
25. Naksuriya, O.; Van Steenbergen, M.J.; Torano, J.S.; Okonogi, S.; Hennink, W.E. A Kinetic Degradation Study of Curcumin in Its Free Form and Loaded in Polymeric Micelles. *AAPS J.* **2016**, *18*, 777–787. [CrossRef] [PubMed]
26. Pardeike, J.; Weber, S.; Zarfl, H.P.; Pagitz, M.; Zimmer, A. Itraconazole-loaded nanostructured lipid carriers (NLC) for pulmonary treatment of aspergillosis in falcons. *Eur. J. Pharm. Biopharm.* **2016**, *108*, 269–276. [CrossRef]
27. Ho, M.-H.; Kuo, P.-Y.; Hsieh, H.-J.; Hsien, T.-Y.; Hou, L.-T.; Lai, J.-Y.; Wang, D.-M. Preparation of porous scaffolds by using freeze-extraction and freeze-gelation methods. *Biomaterials* **2004**, *25*, 129–138. [CrossRef]
28. Kassem, M.A.; Elmeshad, A.N.; Fares, A.R. Enhanced bioavailability of buspirone hydrochloride via cup and core buccal tablets: Formulation and in vitro/in vivo evaluation. *Int. J. Pharm.* **2014**, *463*, 68–80. [CrossRef]
29. Ng, S.-F.; Rouse, J.; Sanderson, D.; Eccleston, G. A Comparative Study of Transmembrane Diffusion and Permeation of Ibuprofen across Synthetic Membranes Using Franz Diffusion Cells. *Pharmaceutics* **2010**, *2*, 209–223. [CrossRef]
30. Wang, D.-P.; Yeh, M.-K. Degradation kinetics of metronidazole in solution. *J. Pharm. Sci.* **1993**, *82*, 95–98. [CrossRef]
31. Bruschi, M.L. 5- Mathematical models of drug release. In *Strategies to Modify the Drug Release from Pharmaceutical Systems*; Bruschi, M.L., Ed.; Elsevier Ltd.: Amsterdam, The Netherlands, 2015; pp. 63–86. [CrossRef]
32. De Caro, V.; Giandalia, G.; Siragusa, M.G.; Sutera, F.M.; Giannola, L.I. New prospective in treatment of Parkinson's disease: Studies on permeation of ropinirole through buccal mucosa. *Int. J. Pharm.* **2012**, *429*, 78–83. [CrossRef]
33. Del Consuelo, I.D.; Pizzolato, G.P.; Falson, F.; Guy, R.H.; Jacques, Y. Evaluation of pig esophageal mucosa as a permeability barrier model for buccal tissue. *J. Pharm. Sci.* **2005**, *94*, 2777–2788. [CrossRef]
34. De Caro, V.; Giandalia, G.; Siragusa, M.G.; Paderni, C.; Campisi, G.; Giannola, L.I. Evaluation of galantamine transbuccal absorption by reconstituted human oral epithelium and porcine tissue as buccal mucosa models: Part I. *Eur. J. Pharm. Biopharm.* **2008**, *70*, 869–873. [CrossRef] [PubMed]
35. Di Prima, G.; Conigliaro, A.; De Caro, V. Mucoadhesive Polymeric Films to Enhance Barbaloin Penetration Into Buccal Mucosa: A Novel Approach to Chemoprevention. *AAPS PharmSciTech* **2019**, *20*, 18. [CrossRef] [PubMed]
36. Shetty, A.V.; Thirugnanam, S.; Dakshinamoorthy, G.; Kajdacsy-balla, A.; Gnanasekar, M. 18α-glycyrrhetinic acid targets prostate cancer cells by down-regulating inflammation-related genes. *Int. J. Oncol.* **2011**, *39*, 635–640. [CrossRef] [PubMed]
37. Damle, M. *Glycyrrhiza glabra* (Liquorice)—A potent medicinal herb. *Int. J. Herb. Med.* **2014**, *2*, 132–136.
38. Trotta, M.; Peira, E.; Debernardi, F.; Gallarate, M. Elastic liposomes for skin delivery of dipotassium glycyrrhizinate. *Int. J. Pharm.* **2002**, *241*, 319–327. [CrossRef]
39. Long, D.R.; Mead, J.; Hendricks, J.M.; Hardy, M.E.; Voyich, J.M. 18β-glycyrrhetinic acid inhibits methicillin-resistant Staphylococcus aureus survival and attenuates virulence gene expression. *Antimicrob. Agents Chemother.* **2013**, *57*, 241–247. [CrossRef]
40. Wang, Y.-J.; Pan, M.-H.; Cheng, A.-L.; Lin, L.-I.; Ho, Y.-S.; Hsieh, C.-Y.; Lin, J.-K. Stability of curcumin in buffer solutions and characterization of its degradation products. *J. Pharm. Biomed. Anal.* **1997**, *15*, 1867–1876. [CrossRef]
41. Brar, S.K.; Verma, M. Measurement of nanoparticles by light-scattering techniques. *TrAC Trends Anal. Chem.* **2011**, *30*, 4–17. [CrossRef]

42. Thatipamula, R.P.; Palem, C.R.; Gannu, R.; Mudragada, S.; Yamsani, M.R. Formulation and in vitro characterization of domperidone loaded solid lipid nanoparticles and nanostructured lipid carriers. *DARU J. Pharm. Sci.* **2011**, *19*, 23–32.
43. Caccavo, D.; Cascone, S.; Lamberti, G.; Barba, A.A.; Larsson, A. Swellable Hydrogel-based Systems for Controlled Drug Delivery. In *Smart Drug Delivery System*; IntechOpen: London, UK, 2016. [CrossRef]

© 2019 by the authors. Licensee MDPI, Basel, Switzerland. This article is an open access article distributed under the terms and conditions of the Creative Commons Attribution (CC BY) license (http://creativecommons.org/licenses/by/4.0/).

Article

Wasted *Ganoderma tsugae* Derived Chitosans for Smear Layer Removal in Endodontic Treatment

Sheng-Tung Huang [1,2], Nai-Chia Teng [3], Hsin-Hui Wang [4], Sung-Chih Hsieh [3,4,*] and Jen-Chang Yang [2,5,6,7,*]

1. Department of Chemical Engineering and Biotechnology, Institute of Biochemical and Biomedical Engineering, National Taipei University of Technology, Taipei 106-08, Taiwan; ws75624@ntut.edu.tw
2. Research Center of Biomedical Device, Taipei Medical University, Taipei 110-52, Taiwan
3. School of Dentistry, College of Oral Medicine, Taipei Medical University, Taipei 110-31, Taiwan; tengnaichia@hotmail.com
4. Department of Dentistry, Taipei Municipal Wan-Fang Hospital, Taipei 116-96, Taiwan; 97240@w.tmu.edu.tw
5. Graduate Institute of Nanomedicine and Medical Engineering, College of Biomedical Engineering, Taipei Medical University, Taipei 110-31, Taiwan
6. International Ph.D. Program in Biomedical Engineering, College of Biomedical Engineering, Taipei Medical University, Taipei 110-31, Taiwan
7. Research Center of Digital Oral Science and Technology, Taipei Medical University, Taipei, 110-52, Taiwan
* Correspondence: endo@tmu.edu.tw (S.-C.H.); Yang820065@tmu.edu.tw (J.-C.Y.); Tel.: +886-2-2736-1661 (ext. 5124) (J.-C.Y.); Fax: 886-2-2736-2295 (S.-C.H.)

Received: 11 October 2019; Accepted: 25 October 2019; Published: 1 November 2019

Abstract: The objective of this study is to investigate the synergistic effects of acid etching and metal-ion chelation in dental smear layer removal using wasted *Ganoderma tsugae* derived chitosans. The wasted *Ganoderma tsugae* fruiting body was used to prepare both acid-soluble fungal chitosan (FCS) and alkali-soluble polysaccharide (ASP). To explore the effective irrigant concentration for smear layer removal, a chelating effect on ferrous ions was conducted. Specimens of various concentrations of EDTA, citric acid, and polysaccharide solutions were reacted with FerroZine™ then the absorbance was examined at 562 nm by a UV-visible spectrophotometer to calculate their metal chelating capability. Twenty extracted premolars were instrumented and individually soaked in the solutions of 15 wt% EDTA, 10 wt% citric acid, 0.04 wt% ASP, 0.04 wt% FCS, and normal saline were randomly divided into five groups (N=4). Next, each tooth was cleaved longitudinally and examined by scanning electron microscopy (SEM) to assay the effectiveness of smear layer removal. The chelating capability for EDTA, FCS, and ASP showed no significant difference over the concentration of 0.04 wt% ($p > 0.05$). The SEM results showed that 0.04 wt% FCS solution was effective in smear layer removal along the canal wall. These results indicated that *Ganoderma tsuage* derived FCS in acid solutions could be a potential alternative as a root canal irrigant solution due to its synergistic effect.

Keywords: EDTA; polysaccharide; chelating effect; smear layer removal

1. Introduction

The goal of endodontic treatment is to clean the root canal system and eliminate the microorganisms as well as necrotic pulp tissue remnants to prevent reinfection of pulpal/periradicular pathosis [1]. Root canal disinfection is usually achieved by mechanical instrumentation and chemical irrigation solution due to the complexity of root anatomy [2]. Smear layer is a thin layer of pulverized mixture comprising of dentin, pulp, and bacterial remnants found spread on root canal walls after instrumentation [3]. Comprehensive literature reviews for smear layer and its clinical implications and relevance to endodontics were reported by Torabinejad et al. [4]. The removal of the smear layer was believed to be beneficial in improving disinfection and good adaptation of dental resin composites to the canal walls. Gu et al. [5] offered an extensive review about irrigant agitation techniques and devices. Many smear layer removal methods were proposed such as chemical [6], ultrasonic [7], and laser [8] techniques, but none of them are totally effective throughout the length of all canals [9].

In operative dentistry, combining ethylenediaminetetraacetic acid (EDTA) with sodium hypochlorite (NaOCl) is the most widely used method in smear layer removal [10]. The chelation of 17 wt% of EDTA (pH 8) with calcium ions is believed to play a major role in promoting decalcification of dentine [11]. However, EDTA is a substituted diamine with cytotoxicity and weak genotoxicity. Oral exposure to EDTA causing adverse reproductive and developmental effects in animals was reported [12]. The 10 wt% citric acid is an alternative EDTA-free irrigant in smear layer removal without any chelating capability reported by Leonardo et al. [13]. It is effective in smear layer removal based on the high solubility of dental hard tissue in acidic environments. To design an irrigant system with high solubility of hydroxyapatite and metal-chelating capability, we proposed the usage of an acid-soluble cleating agent. Among natural polymers, chitosan and its derivatives are known as renewable resources with high chelating capacity and versatile chemical modification [14].

Ganoderma tsugae, known as *Ling-Zhi* in Chinese, is an important fungus with several biological activities in traditional Chinese medicine [15]. The major cell wall content of fungi is chitin (1-4 β-poly(*N*-acetylglucosamine)). Unlike the chitosan produced from crustacean chitin with non-reproducible qualities due to seasonal variation and crustacean species, fungal mycelia are relatively consistent in composition and considered to be free from allergenic animal antigens [16,17]. Currently, the process of alkali-soluble polysaccharide (ASP) prepared from the mild alkaline digesting of *G. tsugae* fruiting body residue is established in our laboratory. Taking advantage of flexibility in preparing acid-soluble fungal chitosan (FCS) and ASP from the same *G. tsugae* fruiting bodies, the decoupling of metal-ion chelation and dissolution mechanisms in smear layer removal becomes possible. The aim of this study was to investigate the possible mechanisms in dental smear layer removal for future optimal irrigant solution formulation.

2. Materials and Methods

G. tsugae residue was obtained from a local mushroom farm in Chiayi County, Taiwan. The stepwise preparation flow diagram of FCS and alkaline deacetylation for ASP are shown in Figure 1a. The reagents for deactylation of *G. tsugae* residues were sodium hydroxide (NaOH, Macron Fine Chemicals, Selangor, Malaysia) and acetic acid (CH_3COOH, Wako Pure Chemical Co., Osaka, Japan). The detailed preparation procedures of FCS were reported elsewhere [18].

The acid-soluble fungal chitosan (FCS) was dissolved in 2% lactic acid at 40 °C for 4 h. Intrinsic viscosities were measured using an automated iVisc Capillary viscometer (Lauda-Königshofen, Germany). The measurement was carried out at 25 °C. The capillary K is 0.009869 mm^2/s^2. Finally, the viscosity average molecular weights of chitosan were calculated using the Mark–Houwink equation.

$$[\eta] = K M v^\alpha \qquad (1)$$

where K of 16.80×10^{-3} mL/g and α of 0.81 were the viscosity parameters used in this calculation [19]. The resulting FCS had a degree of deacetylation (DD) value of 83.7% and molecular weight (MW) of 104 kDa; while ASP revealed DD values of 93.7% and a MW of 43 kDa.

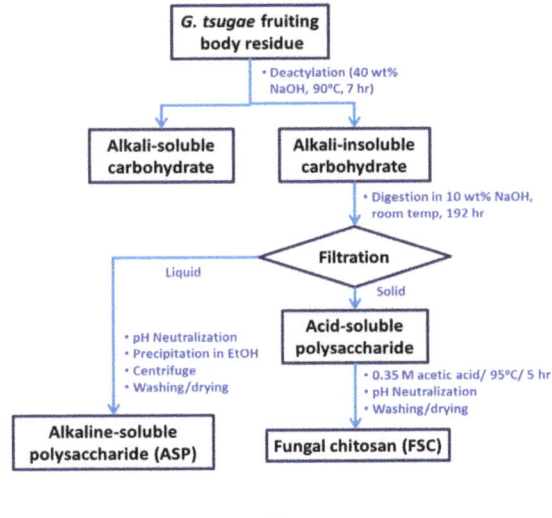

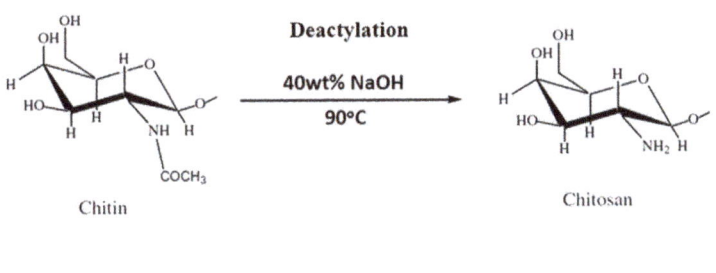

Figure 1. Flow diagram for alkali-soluble polysaccharide (ASP) and fungal chitosan (FSC) preparation: (a) flow diagram of ASP and FSC preparation; (b) deactylation of chitin for preparation of chitosan.

2.1. Chelating Ability

The reagent grade of ferrous chloride (FeCl$_2$·H$_2$O, J.T. Baker, Japan), citric acid (C$_6$H$_8$O$_7$, Sigma-Aldrich, Tokyo, Japan), ethylenediaminetetraacetic acid, disodium salt, dihydrate (Na$_2$EDTA·2H$_2$O) (C$_{10}$H$_{16}$N$_2$O$_8$, J.T. Baker, Chu-Bei City, Taiwan), and FerroZine^{TM} iron reagent (3-(2-pyridyl)-5,6-bis(phenyl sulfonic acid)-1,2,4-triazine, Acros Organics, Gleel, Belgium) for the chelating effect measurement were used as received without purification. Chelating ability was determined according to the method of Riemer et al. [20]. Each chitosan sample (0.1–10 mg/mL) in 2 g/L acetic acid solution (0.2 mL) was mixed with 0.02 mL of 2 mM ferrous chloride. The reaction was initiated by the addition of 0.02 mL of 5 mM FerroZine^{TM}, then shaken vigorously and left at room temperature for 10 min. The absorbance of the mixture was measured by a UV–Visible spectrophotometer (HITACHI, U-2001, Tokyo, Japan) at 562 nm against a blank according to the FerroZine^{TM}-based colorimetric assay. A lower absorbance indicates a higher chelating power. Citric acid and EDTA were used for comparison. Chelating effect was calculated using the following equation:

$$Chelating\,effect\,(\%) = [1 - \frac{OD_{562nm,sample}}{OD_{562nm,blank}}] \times 100 \qquad (2)$$

2.2. Smear Layer Removal

Root canals of access cavities in 20 extracted premolars were prepared by the crown-down technique with rotary nickel-titanium files (K3®; SybronEndo Corporation, Orange, USA) and irrigated with 5 wt% sodium hypochlorite by a 25-gauge blunt-end needle of syringe. The anatomic diameter was determined by introducing successively larger K-files to the working length until resistance was felt upon removal of the file. To investigate the smear layer removal effect of the final irrigation, 20 canals were randomly divided into five groups (N=4) and irrigated with 15 wt% EDTA, 10 wt% citric acids, 0.04 wt% of FSC and ASP solutions, and control (normal saline) for 1 min. All irrigated canals received a final rinse with distilled water for 5 min to halt any chemical activity. The crowns were removed at the cementum–enamel junction. The roots were then split longitudinally with a chisel and a hammer. Finally, specimens were gold-coated and examined by scanning electron microscopy (SEM) (Hitachi S2400, Tokyo, Japan) under an accelerating voltage of 15–20 kV.

2.3. Statistical Analysis

Student's t-test was used to evaluate the statistical significance of the measurement data. The results were considered statistically different at $p < 0.05$.

3. Results

The chelating effect can be quantitatively determined by measuring the rate of color reduction. The absorbance decreased linearly with increasing concentrations of coexisting chelating agents. Figure 2 is the concentration dependence on ferrous ion chelating ability for various irrigant solutions determined colorimetrically. The percentages of chelating activities increased as solution concentration increased. Citric acid showed little chelating ability for ferrous ions within the concentration less than 0.05 wt%. At the concentration of 0.01 wt%, the chelating ability of EDTA, ASP, and FCS were 100%, 52%, and 28%, respectively. The sequence of chelation capability for various solutions was EDTA > ASP > FCS > citric acid when the concentration was less than 0.03 wt%. However, EDTA, ASP, and FCS revealed similar chelation capabilities when the concentration was beyond 0.03 wt%. Based on these findings, the 0.04 wt% of acidic type of FCS solution and basic type of ASP were chosen to compare the smear layer cleaning efficacy to 10 wt% citric acid and 15 wt% EDTA to correlate the effects of the solubility property and chelating capability.

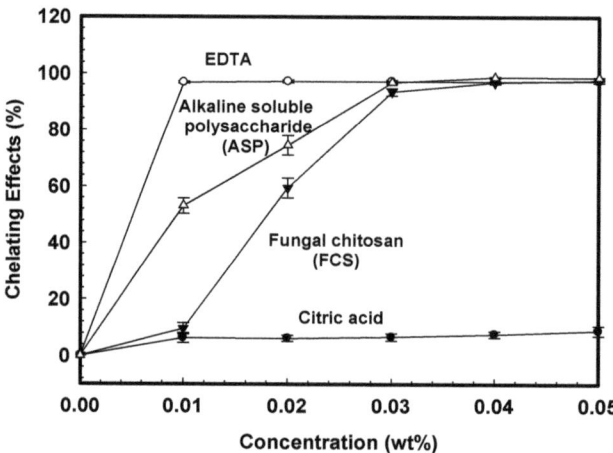

Figure 2. The concentration dependence of chelating effects for various irrigation solutions.

Figure 3 represented the SEM photographs of the root canal surface treated by various irrigation solutions. After treating with 15 wt% EDTA or 10 wt% citric acid, the SEM of middle root canal third (×350) showing most of the smear layer was removed, and some open tubule orifices covered with visible debris of the smear layer on the intertubular and peritubular dentine (Figure 3a,b). Removal of the entire smear layer from dentin surface but with certain severe dentin surface erosion was observed in Figure 3c for the 0.04 wt% FCS solution group. However, the smear layer in the middle third of the root canal was observed after irrigation with 0.04 wt% ASP or normal saline (Figure 3d,e). The efficacy of smear layer removal for 0.04 wt% FCS, 15 wt% EDTA, and 10 wt% citric acid was better than that of 0.04 wt% ASP and normal saline for each concentration indicated in Figure 3.

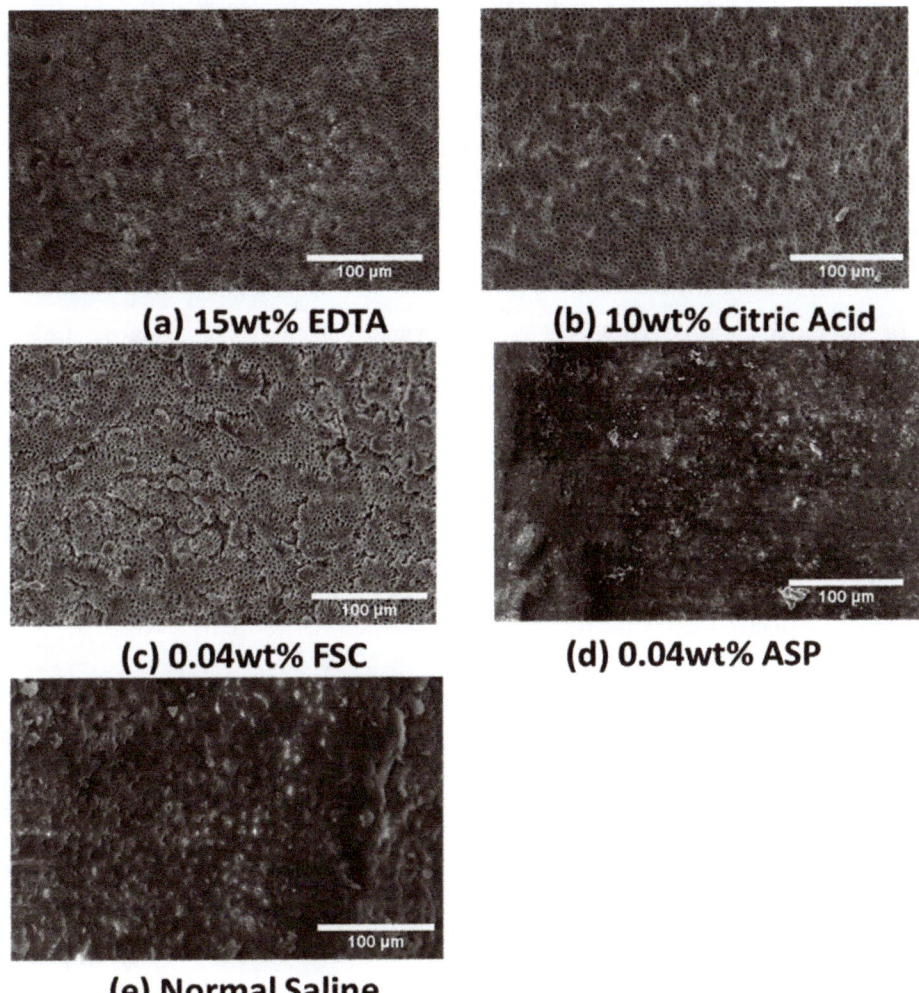

Figure 3. The representative scanning electron microscopy (SEM) photographs (350×) of the prepared surface treated by various irrigation solutions: (**a**) 15 wt% ethylenediaminetetraacetic acid (EDTA); (**b**) 10 wt% citric acid; (**c**) 0.04 wt% fungal chitosan (FCS); (**d**) 0.04 wt% alkali-soluble polysaccharide (ASP); (**e**) normal saline.

4. Discussion

Effective smear-layer removal plays a pivotal role in the successful outcome of endodontic therapy [21]. Due to the similar smear layer cleaning efficacy between 10 wt% citric acid and 15 wt% EDTA, we are interested in knowing what causes the hydroxyapatite (HAp) breakdown shown in SEM.

4.1. Acid Etching Effect

It is important to have a basic understanding about how irrigants interact with HAp. Hydroxyapatite is a crystal with the characteristic of ionic bonds. During the dissolution process, HAp loses its ordered arrangement of the solid mineral phase and becomes free in aqueous solution. Based on the mechanism of the solubility product principle, the water molecules were responsible for disrupting the crystal lattice bonds [22]. In acid dissolution, H^+ ions react with PO_4^{3-} and OH^- ions, which results in reducing their concentration, and then more solid HAp dissolves until the solubility product is re-established, reaching the dissolution equilibrium of HAp [23].

Another mechanism based on a chemical reaction claimed that H^+ ions, not water molecules, were attributed as the primary crystal lattice-disrupting agent [24]. According to the chemical equation of Equation 1, H^+ ions in solution approach the apatite crystal and react with PO_4^{3-} and OH^- ions at the surface of the solid. Conversion to HPO_4^{2-} and H_2O causes a disruption of lattice bonds and release of the ions into solution.

$$Ca_{10}(PO_4)_6(OH)_2 + 8H^+ \rightarrow 10Ca^{2+} + 6HPO_4^{2-} + 2H_2O \tag{3}$$

H^+-induced conformation changes in the collagen matrix might result in contractile stress (ca. 0.2–0.4 MPa) development that was sufficient to cause a collapse of the demineralized dentin matrix [25].

4.2. Chelation Effect

Chelation effect usually involves the formation of multiple coordinate bonds between a multiple bonded ligand (electron pair donor) and a metal ion (electron pair acceptor). EDTA, a hexaprotic weak acid (H_6Y^{2+}) with four carboxylic acids and two ammoniums [26], was first proposed by Nygaard-Ostby to facilitate root canal preparation by chelation [27]. When EDTA is mixed with water, it has an acidic pH, but cannot be used for irrigation as such due to poor solubility. In addition, the action mechanism of EDTA is co-existent with protonation and chelation depending on the pH value as described below [28]:

$$\text{(a) \textbf{Protonation:} } EDTA\text{-}H^{-3} + H^+ \rightarrow EDTA\text{-}H_2^{-2} \text{ (pH} < 7) \tag{4}$$

$$\text{(b) \textbf{Chelation:} } EDTA\text{-}H^{-3} + Ca^{+2} \rightarrow EDTA\text{-}Ca^{-2} + H^+ \text{ (pH} > 7) \tag{5}$$

At a low pH, the protonation of EDTA will reduce the available electro pairs and retard the dissociation of HAp as well as demineralization. On the contrary, the EDTA binding of calcium ions will tend to increase the dissociation of HAp and its availability for chelation at a high or neutral pH. The pH of EDTA has to be adjusted up to a minimum of pH 7 or above to enhance the water solubility but decreases the protonation effect. EDTA chelates calcium ions and forms soluble calcium chelates. When the disodium salt of EDTA is added to this equilibrium, calcium ions are removed from the solution. This leads to the dissolution of further ions from dentin so that the solubility product remains constant.

A continuous rinse with 5 mL of 17% EDTA, as a final rinse for 3 min, efficiently removes the smear layer from root canal walls [29]. But some studies showed that even 1% EDTA solution has a good chelating power [30]. Von der Fehr and Nygaard-Ostby found that EDTA decalcified dentine to a depth of 20 to 30 µm in 5 min [11]. According to Saito et al. greater smear layer removal was found in the 1-min EDTA irrigation group than the 30-sec or 15-sec groups [31]. No clear-cut guideline about the optimal time period for aqueous EDTA solution irrigation was addressed [21]. Prolonged exposure of root dentine to EDTA might have the potential risk of root dentine weakening [32]. Steward et al. found that the combination of EDTA with urea peroxide is very effective in cleaning root canals due to the instrumentation capacity enhancement [33].

4.3. Acid Etching Plus Chelation Effect

Native chitosan is insoluble in water, but its solubility is significantly enhanced when the pH is below its isoelectric point (PI = 6.3) [34]. The chelation of chitosan with metal ions is attributed to the complex formation through the amine groups functioning as ligands [35]. Ion-chelating ability of chitosan is strongly affected by the degree of deacetylation, while fully acetylated chitosan showing little chelating activity [36].

Silva et al. investigated the time-dependent effects of chitosan/acetic acid solution on dentin [37]. Removal of the entire smear layer from the dentin surface was observed with 0.1 wt% chitosan for 3 min, while the severe dentin surface erosion was evidenced by the tubular diameter increasing and extensive destruction of the intertubular dentin in 0.37 wt% chitosan for 5 min. To quantify the efficacy of smear layer removal using chitosan compared with different chelating agents, the concentration of calcium ions in the solutions after irrigation were examined by atomic absorption spectrophotometry with flame [38]. The calcium ion concentration difference between the specimens treated with 15% EDTA and 0.2% chitosan was not significant.

In this study, the group of 0.04 wt% FCS in acetic acid solution revealed complete removal of the entire smear layer from the dentin surface and even with some erosion on the dentin surface showing its high efficiency over 15 wt% EDTA or 10 wt% citric acid. Even with similar chelating ability to FCS, but acid-nonsoluble, 0.04 wt% ASP showed poor smear layer removal outcome. It is evident that both the acid etching and chelation effect are effective for smear layer removal. Furthermore, combining 0.04 wt% FCS in acetic acid solution produces a synergistic effect over the sum of their individual effects.

Like *G. tsugae*, extracts from many herbal/medicinal plants also show antioxidant activity and chelating ability [39]. With the understanding of the interplay between decalcification by metal-ion chelation and acid etching, the measurement of chelating ability offers a good starting point to develop a formulation with more biocompatibility but that is less aggressive to the periapical tissues.

5. Conclusions

The cleaning efficacy of 0.04 wt% FSC was higher than that of 15 wt% EDTA and 10 wt% citric acid. The chitosan dissolved in acid solution offered a synergistic effect of acid etching and chelation on smear layer removal. It suggests that cell wall derivatives from fungal biomasses might provide an alternate irrigant suitable for clinical use.

Author Contributions: S.-C.H. and J.-C.Y. contributed equally to this work. The author contributions listed as following: conceptualization, J.-C.Y.; formal analysis, N.-C.T.; funding acquisition, S.-T.H. and J.-C.Y.; investigation, S.-C.H.; methodology, N.-C.T. and H.-H.W.; project administration, S.-T.H.; resources, S.-C.H.; validation, S.-C.H.; writing—original draft, J.-C.Y.; writing—review and editing, H.-H.W.

Acknowledgments: This work was supported by the grant NTUT-TMU-101-13 from the National Taipei University of Technology–Taipei Medical University (NTUT–TMU) Joint Research Program.

Conflicts of Interest: The authors deny any conflict of interest related to this study.

References

1. Alves, F.R.; Almeida, B.M.; Neves, M.A.; Moreno, J.O.; Rocas, I.N.; Siqueira, J.F., Jr. Disinfecting oval-shaped root canals: Effectiveness of different supplementary approaches. *J. Endod.* **2011**, *37*, 496–501. [CrossRef] [PubMed]
2. Jurič, I.B.; Anić, I. The Use of Lasers in Disinfection and Cleanliness of Root Canals: A Review. *Acta Stomatol. Croat.* **2014**, *48*, 6–15. [CrossRef] [PubMed]
3. McComb, D.; Smith, D.C. A preliminary scanning electron microscopic study of root canals after endodontic procedures. *J. Endod.* **1975**, *1*, 238–242. [CrossRef]

4. Torabinejad, M.; Handysides, R.; Khademi, A.A.; Bakland, L.K. Clinical implications of the smear layer in endodontics: A review. *Oral Surg. Oral Med. Oral Pathol. Oral Radiol. Endod.* **2002**, *94*, 658–666. [CrossRef] [PubMed]
5. Gu, L.S.; Kim, J.R.; Ling, J.; Choi, K.K.; Pashley, D.H.; Tay, F.R. Review of contemporary irrigant agitation techniques and devices. *J. Endod.* **2009**, *35*, 791–804. [CrossRef]
6. Arruda, M.; de Arruda, M.P.; de Carvalho-Junior, J.R.; de Souza-Filho, F.J.; Sousa-Neto, M.D.; de Freitas, G.C. Removal of the smear layer from flattened canals using different chemical substances. *Gen. Dent.* **2007**, *55*, 523–526.
7. Jiang, L.M.; Lak, B.; Eijsvogels, L.M.; Wesselink, P.; van der Sluis, L.W. Comparison of the cleaning efficacy of different final irrigation techniques. *J. Endod.* **2012**, *38*, 838–841. [CrossRef]
8. Harashima, T.; Takeda, F.H.; Kimura, Y.; Matsumoto, K. Effect of Nd:YAG laser irradiation for removal of intracanal debris and smear layer in extracted human teeth. *J. Clin. Laser Med. Surg.* **1997**, *15*, 131–135. [CrossRef]
9. Liu, Y.; Ma, W.; Gao, W. [Effect of different techniques in root canal preparation on coronal microleakage of endodontically treated teeth]. *Hua Xi Kou Qiang Yi Xue Za Zhi* **2012**, *30*, 522–525.
10. Carvalho, A.S.; Camargo, C.H.; Valera, M.C.; Camargo, S.E.; Mancini, M.N. Smear layer removal by auxiliary chemical substances in biomechanical preparation: A scanning electron microscope study. *J. Endod.* **2008**, *34*, 1396–1400. [CrossRef]
11. von der Fehr, F.R.; Nygaard-Östby, B. Effect of edtac and sulfuric acid on root canal dentine. *Oral Surg. Oral Med. Oral Path.* **1963**, *16*, 199–205. [CrossRef]
12. Lanigan, R.S.; Yamarik, T.A. Final Report on the Safety Assessment of EDTA, Calcium Disodium EDTA, Diammonium EDTA, Dipotassium EDTA, Disodium EDTA, TEA-EDTA, Tetrasodium EDTA, Tripotassium EDTA, Trisodium EDTA, HEDTA, and Trisodium HEDTA. *Int. J. Toxicol.* **2002**, *21*, 95–142. [PubMed]
13. Di Lenarda, R.; Cadenaro, M.; Sbaizero, O. Effectiveness of 1 mol L-1 citric acid and 15% EDTA irrigation on smear layer removal. *Int. Endod. J.* **2000**, *33*, 46–52. [CrossRef] [PubMed]
14. Zalloum, H.M.; Mubarak, M.S. Chitosan and chitosan derivatives as chelating agents. In *Natural Polymers, Biopolymers, Biomaterials, and Their Composites, Blends, and IPNs*; Apple Academic Press Inc Point: Pleasant, NJ, USA, 2013.
15. Wei, Y.S.; Wung, B.S.; Lin, Y.C.; Hsieh, C.W. Isolating a cytoprotective compound from Ganoderma tsugae: Effects on induction of Nrf-2-related genes in endothelial cells. *Biosci. Biotechnol. Biochem.* **2009**, *73*, 1757–1763. [CrossRef] [PubMed]
16. Su, C.H.; Sun, C.S.; Juan, S.W.; Ho, H.O.; Hu, C.H.; Sheu, M.T. Development of fungal mycelia as skin substitutes: Effects on wound healing and fibroblast. *Biomaterials* **1999**, *20*, 61–68. [CrossRef]
17. Su, C.H.; Sun, C.S.; Juan, S.W.; Hu, C.H.; Ke, W.T.; Sheu, M.T. Fungal mycelia as the source of chitin and polysaccharides and their applications as skin substitutes. *Biomaterials* **1997**, *18*, 1169–1174. [CrossRef]
18. Chen, C.C.; Cheh, L.W.; Yang, J.C.; Tsai, C.M.; Keh, E.S.; Sheu, M.T. Non-shellfish chitosan from the fruiting body residue of ganoderma tsugae for long-lasting antibacterial guided-tissue regeneration barriers. *J. Dent. Sci.* **2007**, *2*, 19–29.
19. Wang, W.; Bo, S.Q.; Li, S.Q.; Qin, W. Determination of the Mark-Houwink equation for chitosans with different degrees of deacetylation. *Int. J. Biol. Macromol.* **1991**, *13*, 281–285. [CrossRef]
20. Riemer, J.; Hoepken, H.H.; Czerwinska, H.; Robinson, S.R.; Dringen, R. Colorimetric ferrozine-based assay for the quantitation of iron in cultured cells. *Anal. Biochem.* **2004**, *331*, 370–375. [CrossRef]
21. Sudha, R.; Sukumaran, V.R.; Ranganathan, J.; Bharadwaj, N. Comparative evaluation of the effect of two different concentrations of EDTA at two different pH and time periods on root dentin. *J. Conserv. Dent.* **2006**, *9*, 36–42. [CrossRef]
22. Pearce, E.I. On the dissolution of hydroxyapatite in acid solutions. *J. Dent. Res.* **1988**, *67*, 1056–1059. [CrossRef] [PubMed]
23. Griffith, E.J. *Environmental Phosphorus Handbook*; Wiley: New York, NY, USA, 1973.
24. Gray, J.A. Kinetics of the dissolution of human dental enamel in acid. *J. Dent. Res.* **1962**, *41*, 633–645. [CrossRef] [PubMed]
25. Pashley, D.H.; Zhang, Y.; Carvalho, R.M.; Rueggeberg, F.A.; Russell, C.M. H^+-induced tension development in demineralized dentin matrix. *J. Dent. Res.* **2000**, *79*, 1579–1583. [CrossRef] [PubMed]
26. Hargis, L.G. *Analytical Chemistry: Principles and Techniques*; Prentice Hall: Upper Saddle River, NJ, USA, 1988.

27. Nygaard-Östby, B. Chelation in root canal therapy. *Odontol. Tidskr.* **1957**, *65*, 3–11.
28. Pérez, V.C.; Cárdenas, M.E.M.; Planells, U.S. The possible role of pH changes during EDTA demineralization of teeth. *Oral Surg. Oral Med. Oral Pathol.* **1989**, *68*, 220–222. [CrossRef]
29. Mello, I.; Kammerer, B.A.; Yoshimoto, D.; Macedo, M.C.; Antoniazzi, J.H. Influence of final rinse technique on ability of ethylenediaminetetraacetic acid of removing smear layer. *J. Endod.* **2010**, *36*, 512–514. [CrossRef]
30. Zehnder, M. Root canal irrigants. *J. Endod.* **2006**, *32*, 389–398. [CrossRef]
31. Saito, K.; Webb, T.D.; Imamura, G.M.; Goodell, G.G. Effect of shortened irrigation times with 17% ethylene diamine tetra-acetic acid on smear layer removal after rotary canal instrumentation. *J. Endod.* **2008**, *34*, 1011–1014. [CrossRef]
32. Calt, S.; Serper, A. Time-dependent effects of EDTA on dentin structures. *J. Endod.* **2002**, *28*, 17–19. [CrossRef]
33. Stewart, G.G.; Kapsimalas, P.; Rappaport, H. EDTA and urea peroxide for root canal preparation. *J. Am. Dent. Assoc.* **1969**, *78*, 335–338. [CrossRef]
34. Kozulic, B. Looking at bands from another side. *Anal. Biochem.* **1994**, *216*, 253–261. [CrossRef] [PubMed]
35. Roberts, G.A.F. *Chitin Chemistry*; Macmillan: London, UK, 1992.
36. Qin, Y. The chelating properties of chitosan fibers. *J. Appl. Polym. Sci.* **1993**, *49*, 727–731. [CrossRef]
37. Silva, P.V.; Guedes, D.F.; Pecora, J.D.; da Cruz-Filho, A.M. Time-dependent effects of chitosan on dentin structures. *Braz. Dent. J.* **2012**, *23*, 357–361. [CrossRef] [PubMed]
38. Silva, P.V.; Guedes, D.F.; Nakadi, F.V.; Pecora, J.D.; Cruz-Filho, A.M. Chitosan: A new solution for removal of smear layer after root canal instrumentation. *Int. Endod. J.* **2013**, *46*, 332–338. [CrossRef] [PubMed]
39. Gutiérrez, R.M.P. *Handbook of Naturally Occurring Compounds with Antioxidant Activity in Plants*; Nova Science: New York, NY, USA, 2006.

© 2019 by the authors. Licensee MDPI, Basel, Switzerland. This article is an open access article distributed under the terms and conditions of the Creative Commons Attribution (CC BY) license (http://creativecommons.org/licenses/by/4.0/).

Article

The Development of Gelatin/Hyaluronate Copolymer Mixed with Calcium Sulfate, Hydroxyapatite, and Stromal-Cell-Derived Factor-1 for Bone Regeneration Enhancement

Yun-Liang Chang [1,2], Chia-Ying Hsieh [1], Chao-Yuan Yeh [3] and Feng-Huei Lin [1,*]

1. Department of Biomedical Engineering, National Taiwan University, No. 1, Sec.1, Jen-Ai Road, Taipei City 10051, Taiwan
2. Department of Orthopaedic Surgery, National Taiwan University Hospital, No. 7, Chung Shan South Road, Taipei City 10002, Taiwan
3. Integrative Stem Cell Center, China Medical University, No. 2, Yude Road, Taichung City 40447, Taiwan
* Correspondence: double@ntu.edu.tw; Tel.: +886-2-2732-0443

Received: 8 August 2019; Accepted: 2 September 2019; Published: 5 September 2019

Abstract: In clinical practice, bone defects still remain a challenge. In recent years, apart from the osteoconductivity that most bone void fillers already provide, osteoinductivity has also been emphasized to promote bone healing. Stromal-cell-derived factor-1 (SDF-1) has been shown to have the ability to recruit mesenchymal stem cells (MSCs), which play an important role in the bone regeneration process. In this study, we developed a gelatin–hyaluronate (Gel-HA) copolymer mixed with calcium sulfate (CS), hydroxyapatite (HAP), and SDF-1 in order to enhance bone regeneration in a bone defect model. The composites were tested in vitro for biocompatibility and their ability to recruit MSCs after material characterization. For the in vivo test, a rat femoral condyle bone defect model was used. Micro computed tomography (Micro-CT), two-photon excitation microscopy, and histology analysis were performed to assess bone regeneration. As expected, enhanced bone regeneration was well observed in the group filled with Gel-HA/CS/HAP/SDF-1 composites compared with the control group in our animal model. Furthermore, detailed blood analysis of rats showed no obvious systemic toxicity or side effects after material implantation. In conclusion, the Gel-HA/CS/HAP/SDF-1 composite may be a safe and applicable material to enhance bone regeneration in bone defects.

Keywords: SDF-1; bone defect; bone regeneration; mesenchymal stem cells; gelatin; hyaluronate; calcium sulfate; hydroxyapatite; biomaterial

1. Introduction

Bone defect is a common clinical scenario during treatment of bone-related pathology. Comminuted fractures, neoplasms, and bone infections are common causes of bone defect [1]. Apart from autografts or allografts, different kinds of synthetic bone substitutes have been used by clinical doctors for decades to treat bone defects [2–4]. However, each kind of material still has its own limitations.

Calcium sulfate (CS) has been used to treat bone defects since 1892 and is still presently useful for different pathologies [5,6]. CS is cheap, can be synthesized easily, and has osteoconductivity and good mechanical strength [7]. However, the rapid resorption of CS may cause some problems, including serious wound discharge or delayed bone union [8,9]. Hydroxyapatite (HAP) and tricalcium phosphate (TCP) are both osteoconductive and have been widely used for bone defects [10,11]. Hyaluronate (HA) has a relatively poor resorption rate compared with CS, and TCP falls in between. It takes 6–18 months for TCP and up to 10 years for HAP to be completely resorbed in human beings. However, due to the lack of osteoinductivity, all these ceramics are not suitable for use in areas with poor vascular perfusion or large segmental defects.

Bone morphogenetic proteins (BMPs) such as BMP-2 or BMP-7 are also used to treat bone defects, as their osteoinductivity can promote bone repair [2–4,12]. However, BMPs are very expensive and have some pro-oncogenic concerns. Demineralized bone matrix (DBM) has become increasingly popular due to its osteoinductivity and the various forms it can take for different clinical needs [13,14]. DBM is made from allograft with the growth factors remaining, and it is suitable for filling bone defects due to its collagen scaffold [15,16]. It is also slightly cheaper than BMPs and has no pro-oncogenic concerns. However, the potential for disease transmission is still a problem due to its allograft-derived nature.

In the management of bone defects, the osteoconductivity and osteoinductivity of bone void fillers are equally important. Using a synthetic bone substitute seems to be a better way to avoid potential disease transmission, but most bone substitutes lack osteoinductivity. Stromal-cell-derived factor-1 (SDF-1) is a dominant chemokine in bone marrow which can be induced in the periosteum of injured bone [17]. In addition to many signaling functions in different organs, it is able to recruit mesenchymal stem cells (MSCs) [18,19]. Due to the potential benefit of MSCs in bone regeneration, SDF-1 may be considered to have a certain degree of osteoinductive effect, resulting in enhanced bone growth and new bone formation [20,21]. Gelatin (Gel) is a denatured collagen and hyaluronate (HA) is a non-sulfated glycosaminoglycan. Both of them have good biodegradability and biocompatibility. Gel can enhance cell proliferation, differentiation and attachment, while HA may influence cell mobility, cell–matrix adhesion and cell–cell interaction [22,23]. By mixing Gel and HA together, it can create a biocompatible and biodegradable scaffold to contain various bioactive agents or materials. In this study, we integrated a gelatin–hyaluronate (Gel-HA) copolymer with CS and HAP particles after being crosslinked by 1,4-butanediol diglycidyl ether (BDDE), as these provide good mechanical properties as bone fillers and drug carriers. SDF-1 was also added in order to enhance MSC recruitment and to test the possible bone growing effect. We hypothesized that this Gel-HA/CS/HAP composite with SDF-1 may promote bone healing in a bone defect animal model.

2. Materials and Methods

2.1. Materials

Several materials were purchased from Sigma-Aldrich (St. Louis, MO, USA), including gelatin type A from porcine skin (Gel), hyaluronic acid sodium salt (HA), BDDE, calcium sulfate hemihydrate (CS), and calcium hydroxide. SDF-1 was purchased from PeproTech (Rocky Hill, NJ, USA). Human umbilical cord blood mesenchymal stem cells immortalized by human telomerase reverse transcriptase (cbMSC-hTERT)s were obtained from the American Type Culture Collection (ATCC).

2.2. Preparation of Hydroxyapatite

To obtain the optimal particle size, hydroxyapatite was made in our own laboratory. Phosphoric acid (0.3 M) was added dropwise to a 0.5 M calcium hydroxide solution at the rate of 3 mL/min, and the pH value was adjusted to 8.5. The mixture was then stirred for 2 h at 85 °C and left standing at 85 °C for 24 h. After that, the precipitated powder was collected. The powder was then washed three times with double-distilled water and freeze-dried [24,25].

2.3. Preparation of Gel-HA/CS/HAP Composite with SDF-1

First, Gel (10 wt %, 300 Bloom) and HA (0.5 wt %) were dissolved in distilled water above 37 °C, respectively. At the volume ratio of 85:15, the Gel solution and HA solution were well mixed. In the meantime, BDDE (0.5 vol %) was added to the mixture as a crosslinker. Using a magnetic stirrer, the mixture was mixed for 24 h at 37 °C. After that, CS and HAP were mixed at the ratio of 50:50 as an inorganic mixture. At the weight ratio of 75:25, the crosslinked organic mixture (Gel and HA) and inorganic mixture (CS and HAP) were mixed. Then, 100 ng/mL of SDF-1 was added to the mixture after the temperature of the mixture dropped to room temperature. The final mixture was then poured into a petri dish and kept at 4 °C.

2.4. Material Characterization

2.4.1. X-ray Diffraction (XRD) Analysis

X-ray diffraction (Rigaku, TTRAX 3, Tokyo, Japan) was used to characterize the composition of the Gel-HA/CS/HAP composite. The tension of the XRD was set to 30 kV, and the current was 20 mA. The scanning rate was 10°/min, while the scanning degree was set to $2\theta = 10°–80°$.

2.4.2. Fourier-Transform Infrared Spectroscopy Analysis

Using Fourier-transform infrared spectroscopy (Jasco, FT/IR-4200, Tokyo, Japan), the success of Gel and HA crosslinking with BDDE was determined.

2.4.3. Swelling Ratio

At 37 °C, the *swelling ratio* of the Gel-HA/CS/HAP composite was determined by incubation in phosphate buffer saline (PBS, pH 7.4). The composite with a known weight (W_d) was placed in PBS for various time points. After the surface-adsorbed water was removed by filter paper, the wet weight (W_s) of the composite was obtained. The *swelling ratio* was calculated as follows:

$$\text{Swelling ratio (\%)} = \frac{W_s - W_d}{W_d} \times 100\% \tag{1}$$

2.4.4. Degradation Test

For the in vitro degradation test, the Gel-HA/CS/HAP composite was immersed in PBS solution (pH 7.4) at 37 °C. The degraded gelatin solution was measured by an ELISA reader (Tecan, Sunrise, Melbourne, Australia) at 230 nm at various time points.

2.4.5. SDF-1 Release Profile

Considering the cost of SDF-1, we chose fluorescein isothiocyanate (FITC) conjugated insulin as the model drug. Since insulin (6 kDa) shares a similar molecular weight and shape with SDF-1 (8 kDa), it may also present a similar release profile to SDF-1. The Gel-HA/CS/HAP composite with FITC-insulin was immersed in PBS solution (pH 7.4) at 37 °C. Then, the released FITC–insulin solution was measured by an ELISA reader at 488 nm at various time points.

2.5. In Vitro Study

2.5.1. Cell Culture

To culture the cbMSC-hTERT cells, α-minimum essential medium (MEM) supplemented with 20% fetal bovine serum (FBS), 1% penicillin/streptomycin/amphotericin B, 1.0 mM sodium pyruvate, 4 ng/mL recombinant human basic fibroblast growth factor (rHubFGF), and 30 µg/mL hygromycin was used. Under an atmosphere of 5% CO_2, cbMSC-hTERT cells were culture in a humidified incubator at 37 °C [26].

2.5.2. Cell Viability

To test cell viability, cbMSC-hTERT cells (10^4 cells/well) were cultured on a 96-well plate for 24 h. Then, 100 µL of the material extracts (0.2 g material/mL media) were added into the wells and incubated for 24 h. After that, 100 µL of water-soluble tetrazolium salt (WST-1) solution was added to the wells and incubated for 2 h. Using an ELISA reader, the absorbance values of each well were measured at 450 nm.

2.5.3. MSC Recruitment Test

cbMSC-hTERT cells (10^6 cells/well) were cultured on a six-well plate for 24 h. Using cell scrapers, half of the cells were then removed. In the area without cells, the Gel-HA/CS/HAP composites with or without SDF-1 were prepared and placed into the six-well plate. The migration of cbMSC-hTERT cells was observed at 0, 24, 48, and 72 h.

2.6. In Vivo Study

2.6.1. Implantation of Composite in Rat Femur Bone Defect Model

Eighteen Wistar male rats (300 g) were purchased commercially and kept in different cages in groups of three. Water and food were provided properly. To acclimate them to the environment, these animals were first kept a week in an animal house. Before operation, the rats were anesthetized with isoflurane. The fur around the left lower limb was gently shaved, and the skin was disinfected with alcohol. A midline skin incision was made over the left knee, and the medial femoral condyle was exposed. A bone defect that was 2.5 mm in diameter and 5 mm in depth was created over the medial femoral condyle by a trephine drill [27–31]. The Gel-HA/CS/HAP composite without SDF-1 and the composite with 100 ng/mL of SDF-1 were prepared and filled into the bone defect based on the previously assigned group. The bone defect was left in situ without any implant in the control group. The incision was then closed with 4–0 sutures. The animals were allowed to recover and were kept under proper care after operation. Three rats of each group were sacrificed one or two months after implantation. The whole femur bone specimen was collected gently. All animal studies were performed according to the protocol approved by the Institutional Animal Care and Use Committee (IACUC) of the National Taiwan University College of Medicine and College of Public Health.

2.6.2. Blood Analysis

Before sacrifice, a cardiac puncture was performed with a 23G needle under anesthesia. The obtained blood sample was divided between two different tubes. One tube was centrifuged, and serum was extracted for biochemical analysis. The other tube contained dipotassium ethylenediaminetetraacetic acid (K2EDTA), which was sent for whole blood analysis. For the biochemical test, lactate dehydrogenase (LDH), alkaline phosphatase (ALKP), and Ca were measured. For the whole blood test, red blood cells (RBCs), hemoglobulin (HGB), hematocrit (HCT), mean corpuscular volume (MCV), mean corpuscular hemoglobin (MCH), mean corpuscular hemoglobin concentration (MCHC), white blood cells (WBCs), neutrophil (NEUT), lymphocyte (LYMPH), monocyte (MONO), eosinophil (EO), and basophil (BASO) were measured.

2.6.3. Two-Photon Excitation Microscopy

Right after sacrifice, fresh femur specimens were inspected under two-photon excitation microscopy. Using an 890 nm laser beam as the excitation light source, the microscope received a double frequency of 445 ± 10 nm to generate a signal to observe type I collagen of bone tissue and the surrounding soft tissue. At the same time, it received a spontaneous fluorescent signal of 500–530 nm from the specimens.

2.6.4. Micro-CT

For the micro-CT analysis, the obtained femur bone specimens were fixed in 10% formaldehyde for a week and then transferred to 95% ethanol. The distal part of the femur was scanned by a Bruker SkyScan 1076 micro-CT. The reconstructed micro-CT data were analyzed to determine the amount of bone and the presence of the implants and to calculate the bone volume/tissue volume (BV/TV). DataViewer and CTAn softwares (Bruker, Billerica, Massachusetts, USA) were used to make 2D and 3D images, respectively.

2.6.5. Histological Analysis

Femur specimens were first fixed with 10% formaldehyde for a week and then decalcified with 5% nitric acid for three days. After decalcification, the specimens were dehydrated through a sequential alcohol (70–100%) treatment. The dehydrated specimens were then transitioned to xylene series and eventually paraffin for embedding. Paraffin-embedded samples were then cut into 5-μm sections and attached to glass slides. The sections were stained with hematoxylin and eosin (H&E) and Masson's trichrome (MT) to visualize the dense tissue and new bone formation.

3. Results and Discussion

3.1. XRD Analysis

The XRD pattern of the Gel-HA/CS/HAP composite is shown in Figure 1. Comparing the patterns, the Gel/HA/CS/HAP composite (Figure 1g) had a composition similar to the human femur bone (Figure 1a). There was a broad peak at 2θ = ~20° among the patterns of sodium hyaluronate (Figure 1b), gelatin (Figure 1c), and the Gel-HA/CS/HAP composite (Figure 1g). This characteristic peak was assigned to the crystalline structure in hyaluronic acid and the triple-helical crystalline structure in gelatin. Comparing the patterns of calcium sulfate (Figure 1d) and the Gel-HA/CS/HAP composite (Figure 1g), it was confirmed that the composite did contain calcium sulfate, which did not change in structure after the crosslinking process. Furthermore, comparing the pattern of the hydroxyapatite prepared in this study (Figure 1e) and the standard pattern provided by the Joint Committee on Powder Diffraction Standards (JCPDS) (Figure 1f), the prepared hydroxyapatite was consistent with the standard, and the crosslinking process also did not affect the structure of the hydroxyapatite.

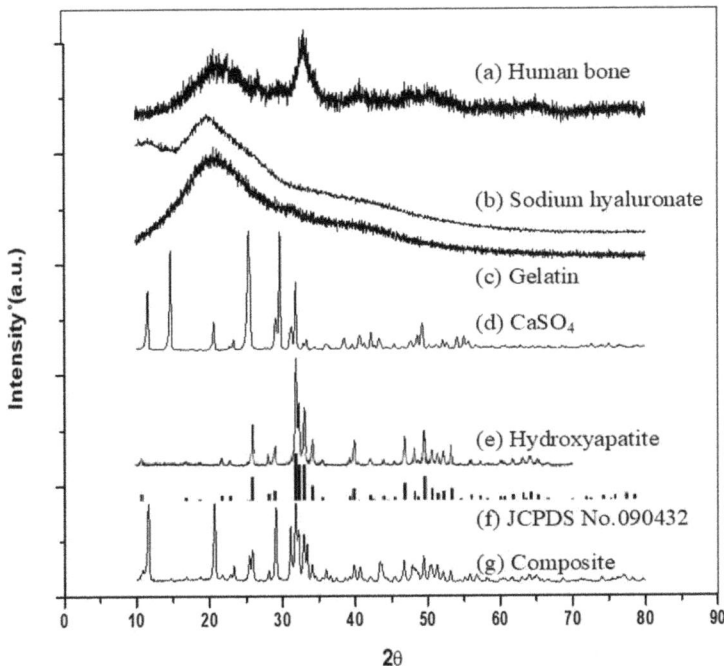

Figure 1. X-ray diffraction (XRD) patterns: (**a**) human bone, (**b**) sodium hyaluronate, (**c**) gelatin, (**d**) calcium sulfate, (**e**) hydroxyapatite, (**f**) Joint Committee on Powder Diffraction Standards (JCPDS) no. 090432, and (**g**) gelatin–hyaluronate/calcium sulfate/hydroxyapatite (Gel-HA/CS/HAP) composite.

3.2. Fourier-Transform Infrared Spectroscopy (FTIR) Analysis

The results of the FTIR analysis are shown in Figure 2. On the FTIR spectrum of the Gel-HA/CS/HAP composite (Figure 2a) and calcium sulfate (Figure 2b), the absorbance bands of S–O bonding were detected at 1151, 1148, 1098, 1095, 674, and 673 cm^{-1}. For the spectrum of hydroxyapatite (Figure 2c), 1091, 1030, 963, 631, 601, and 564 cm^{-1} were the absorbance bands of P–O bonding. Among the FTIR spectra of the composite, sodium hyaluronate (Figure 2d), and gelatin (Figure 2e), the absorbance band of amide I C = O bonding could be seen at 1637–1639 cm^{-1}, while the absorbance band of amide II N–H and C–N bonding were found at 1520, 1542, and 1537 cm^{-1}. For the composite and sodium hyaluronate, 1151 and 1159 cm^{-1} were the absorbance bands of ether C–O–C bonding. The absorbance band of alcohol group C–OH bonding could be seen on the FTIR spectrum of sodium hyaluronate at 1030 cm^{-1}. As for the absorbance band of amide III C–N and N–H bonding, it was shown in the spectrum of the composite and gelatin at 1235 cm^{-1}.

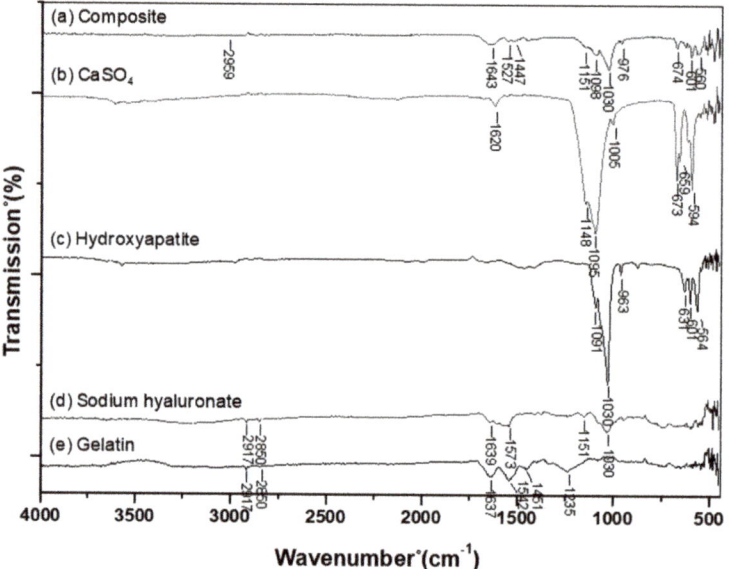

Figure 2. Fourier-transform infrared spectroscopy (FTIR) spectrum: (**a**) Gel-HA/CS/HAP composite, (**b**) calcium sulfate, (**c**) hydroxyapatite, (**d**) sodium hyaluronate, and (**e**) gelatin. The absorbance band at 2959 cm^{-1} was the C–H bonding stretching of 1,4-butanediol diglycidyl ether (BDDE), indicating that the gelatin and hyaluronic acid in the composite of this study were successfully crosslinked.

As previously mentioned, the FTIR spectrum was used to determine the success of crosslinking. Many ether bonds were formed when HA and Gel were crosslinked by BDDE. However, in the HA backbone, there were many ether bonds, so it was not possible to use the absorbance of an ether bond to determine whether the crosslinking process was successful. From the FTIR results of the composite, there was an absorbance band at 2959 cm^{-1}. This absorbance band was the C–H bonding stretching of BDDE, which indicated that the Gel-HA/CS/HAP composite was successfully crosslinked.

3.3. Swelling Ratio

As is well known, the swelling of a composite is caused by the absorption of water. According to the results shown in Figure 3, the Gel-HA/CS/HAP composite underwent two-stage swelling. The composite used in this study was copolymerized by Gel and HA, which naturally have different water absorption abilities. Furthermore, Gel conducted less water absorption after being crosslinked. Therefore, the first stage of swelling was attributed to HA, for which water absorption was faster. The second stage of swelling was attributed to Gel since HA already attained its maximum water absorption.

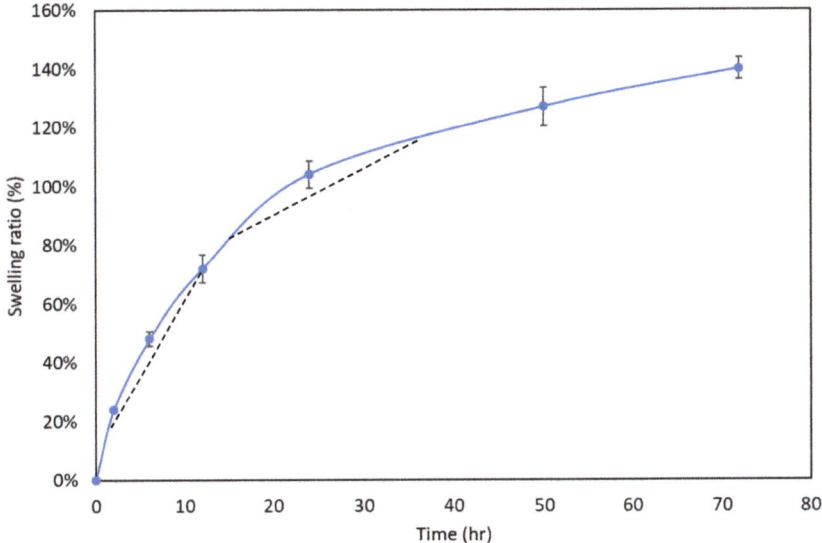

Figure 3. Swelling ratio test (n = 6). The Gel-HA/CS/HAP composite performed two-stage swelling. The first stage was the result of hyaluronic acid, and the second stage was caused by gelatin.

3.4. Degradation Rate

Degradation occurred after the composite absorbed water and swelled. In our study, the Gel-HA/CS/HAP composite exhibited two-stage degradation (Figure 4). From the results of the swelling ratio analysis, Gel initially contributed less to swelling. Therefore, Gel also expressed less degradation at the first stage of composite degradation. After Gel started to swell dramatically, it considerably degraded during the second stage of degradation.

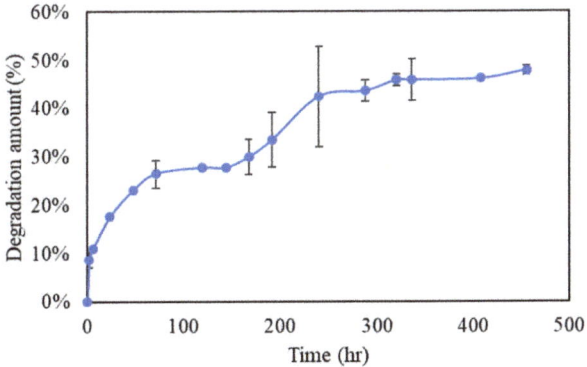

Figure 4. In vitro degradation test (n = 6). The Gel-HA/CS/HAP composite underwent two-stage degradation. At the first stage, due to the fact that gelatin did not swell dramatically, little degradation occurred. At the second stage, gelatin absorbed water and swelled considerably, causing a relatively large amount of degradation.

3.5. SDF-1 Release Profile

The release profile of SDF-1 was also divided into two stages (Figure 5). Since SDF-1 was not crosslinked with Gel or HA, the first-stage release was probably caused by SDF-1, which adsorbed at the surface of the Gel-HA/CS/HAP composite. After that, the second-stage release lasted for nearly five days. It was relatively slow and stable, which may have been the result of the degradation of the Gel-HA/CS/HAP composite—mostly the degradation of Gel, followed by the release of SDF-1. According to previous studies, collagen has a binding domain for SDF-1 [32–34]. As is known, HA is a polysaccharide and Gel is a denatured collagen. Therefore, Gel might have a higher affinity to SDF-1 than HA. The course of SDF-1 release lasted for five days in total, and the effective release was about 90%.

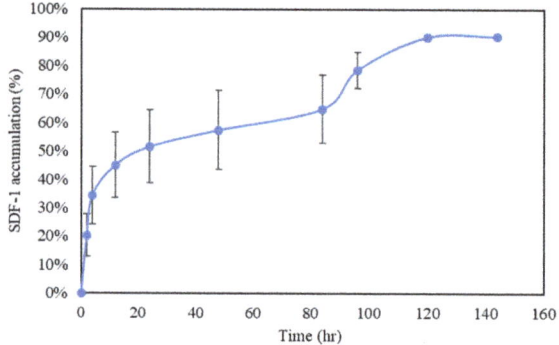

Figure 5. Stromal-cell-derived factor-1 (SDF-1) release profile (n = 6). The Gel-HA/CS/HAP underwent a two-stage release. The first stage of release was caused by the SDF-1 molecules at the surface, and the second stage of release was caused by the degradation of the composite.

3.6. Cell Viability

Cell viability and proliferation were determined by a WST-1 assay. cbMSC-hTERT cells were treated with the medium extracts of the negative control (aluminum oxide, Al_2O_3), positive control (zinc diethyldithiocarbamate, ZDEC), and the Gel-HA/CS/HAP composite, respectively, at the concentration of 0.2 g/mL. The control group (nontreated) was defined as 100% cell viability. As expected, the negative control and Gel-HA/CS/HAP composite did not affect the growth of cbMSC-hTERT cells (Figure 6). However, the positive control did inhibit the growth of cbMSC-hTERT cells.

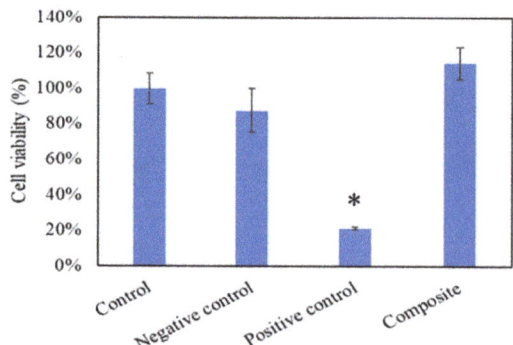

Figure 6. WST-1 test (n = 6). Control group was 100% cell viability, negative control was aluminum oxide, and positive control was zinc diethyldithiocarbamate. The results showed that the Gel-HA/CS/HAP composite had no cell toxicity. * $p < 0.001$.

3.7. MSC Recruitment Test

In order to confirm the SDF-1 chemotaxis ability of the composite, an MSC recruitment test was conducted. cbMSC-hTERT cells were placed at the right and the material was placed at the left in the beginning (0 h) in all groups (Figure 7). After 24 h, the cbMSC-hTERT cells of the control group and the group without SDF-1 started to form colonies due to its fibroblast-like nature. However, the cbMSC-hTERT cells did show some cell migration from right to left in the 100 ng/mL SDF-1 group, which was related to the SDF-1 concentration gradient released from the Gel-HA/CS/HAP composite on the left side. After 48 and 72 h, the phenomena of each group became more apparent. According to the cell-number–time analysis, cbMSC-hTERT cells moved toward the Gel-HA/CS/HAP/SDF-1 composite as the time increased (Figure 8a). From the cell-number–distance analysis, cbMSC-hTERT cells migrated toward the Gel-HA/CS/HAP/SDF-1 composite as the distance increased from the left side (Figure 8b). All these results indicated that the Gel-HA/CS/HAP/SDF-1 composite was capable of attracting MSCs. Moreover, SDF-1 did not change the fibroblast-like type of MSCs or have any impact on MSCs.

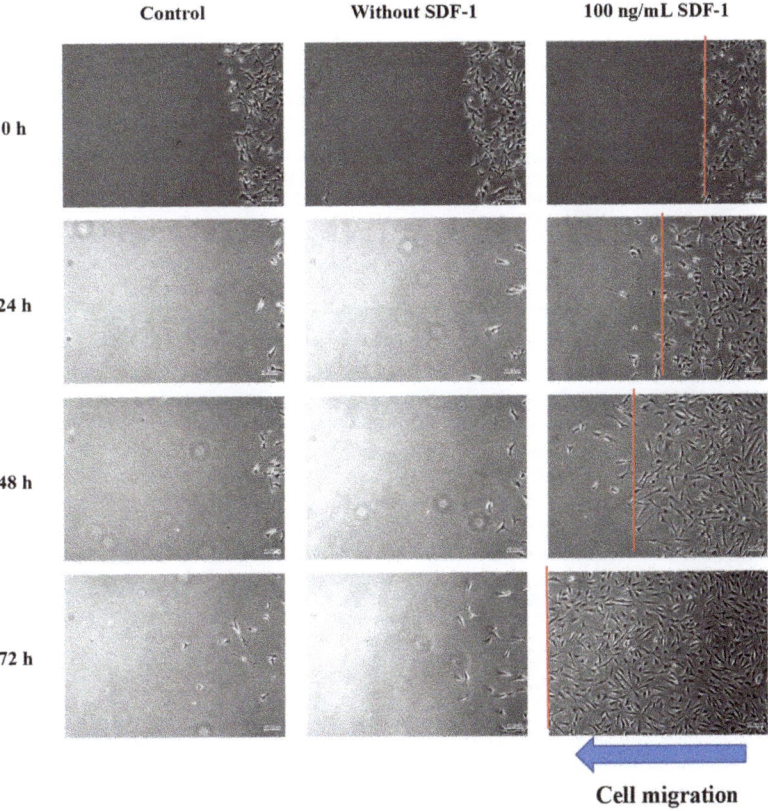

Figure 7. Mesenchymal stem cell (MSC) recruitment test (40×). First, second, and third columns show MSC migration of the control group, without SDF-1 group, and the 100 ng/mL SDF-1 group, respectively. The red line indicates the cell front. The results show that the MSCs of the control group and without SDF-1 group gathered together and formed colonies. MSCs of the 100 ng/mL group migrated toward the concentration gradient of SDF-1.

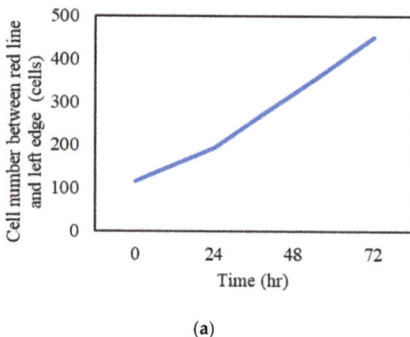 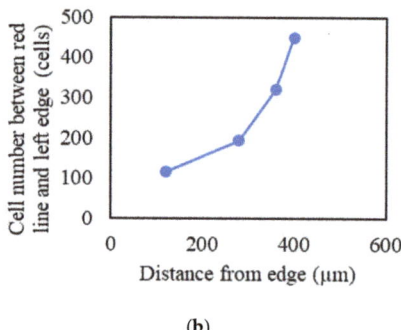

Figure 8. MSC recruitment test analysis: (**a**) cell number between the red line and left edge versus time and (**b**) cell number between the red line and left edge versus distance from the left edge.

3.8. In Vivo Study

In the in vivo study, no infection or severe side effects were observed at implant sites and no deaths were recorded during the recovery period. Soft tissue wounds recovered without showing any acute inflammation in experimental, control, and comparison groups.

3.8.1. Blood Analysis

Before sacrifice, cardiac puncture was done for biochemical and whole blood analyses, and the results were compared with standard range (Charles River Laboratories, 1982). The biochemistry tests after one and two months are shown in Tables 1 and 2. As we can see, the Ca values of all groups were in a normal range. LDH values of the 100 ng/mL SDF-1 group after two months were slightly higher than the normal range but were still acceptable. The results of the whole blood tests are shown in Tables 3 and 4. The WBC values of all groups were elevated slightly after one month but returned to normal after two months, which may have been related to the inflammatory response caused by the created bone defects. Based on the results of the biochemical and whole blood analyses, the Gel-HA/CS/HAP composites with or without SDF-1 prepared in this study appeared to have no obvious systemic toxicity.

Table 1. Biochemistry test (one month).

	Control	Without SDF-1	With SDF-1	Reference *
ALKP (U/L)	310.67	276	253	39–216
Ca (mg/dL)	11.63	12.1	12.27	8–15
LDH (U/L)	656.33	459.33	430	300–700

* Charles River Laboratories, 1982. ALKP: alkaline phosphatase; LDH: lactate dehydrogenase.

Table 2. Biochemistry test (two months).

	Control	Without SDF-1	With SDF-1	Reference *
ALKP (U/L)	260.33	214.67	214.67	39–216
Ca (mg/dL)	10.97	10.77	10.83	8–15
LDH (U/L)	586	368.67	771	300–700

* Charles River Laboratories, 1982. ALKP: alkaline phosphatase; LDH: lactate dehydrogenase.

Table 3. Whole blood test (one month).

	Control	Without SDF-1	With SDF-1	Reference *
RBC (M/µL)	8.32	8.45	8.3	7.37–9.25
HGB (g/dL)	14.73	15.3	15.1	14.4–17.6
HCT (%)	44.4	46.37	45.63	36–46
MCV (fL)	53.37	54.83	54.97	47–52
MCH (pg)	17.7	18.1	18.2	17–21
MCHC (g/dL)	33.2	33	33.1	35–43
WBC (K/µL)	11.52	13.99	12.33	6.19–12.55
NEUT (%)	16.67	14.13	19.23	1–29
LYMPH (%)	75.27	80.9	73.57	70–99
MONO (%)	5.23	2.97	5.13	0–6
EO (%)	2.63	1.9	2	0–3
BASO (%)	0.2	0.1	0.07	0–2

* Charles River Laboratories, 1982. RBC: red blood cell; HGB: hemoglobulin; HCT: hematocrit; MCV: mean corpuscular volume: MCH: mean corpuscular hemoglobin; MCHC: mean corpuscular hemoglobin concentration; WBC: white blood cell; NEUT: neutrophil; LYMPTH: lymphocyte; MONO: monocyte; EO: eosinophil; BASO: basophil.

Table 4. Whole blood test (two months).

	Control	Without SDF-1	With SDF-1	Reference *
RBC (M/µL)	9.16	9.18	8.87	7.37–9.25
HGB (g/dL)	15.47	15.4	15.00	14.4–17.6
HCT (%)	45	45.47	43.75	36–46
MCV (fL)	49.14	49.58	49.35	47–52
MCH (pg)	16.90	16.79	16.92	17–21
MCHC (g/dL)	34.37	33.89	34.29	35–43
WBC (K/µL)	10.29	10.33	10.05	6.19–12.55
NEUT (%)	16.67	14.13	19.23	1–29
LYMPH (%)	75.27	80.9	73.57	70–99
MONO (%)	5.23	2.97	5.13	0–6
EO (%)	2.63	1.9	2	0–3
BASO (%)	0.2	0.1	0.07	0–2

* Charles River Laboratories, 1982. RBC: red blood cell; HGB: hemoglobulin; HCT: hematocrit; MCV: mean corpuscular volume: MCH: mean corpuscular hemoglobin; MCHC: mean corpuscular hemoglobin concentration; WBC: white blood cell; NEUT: neutrophil; LYMPTH: lymphocyte; MONO: monocyte; EO: eosinophil; BASO: basophil.

3.8.2. Two-Photon Excitation Microscopy

Second harmonic generation (SHG) is an optical phenomenon in which two incident photons are converted into a single photon by interacting with the crystalline optical structure of some materials. In biological tissues, type I collagen is known to have strong SHG signals. We used a two-photon microscope with 840 nm excitation and a 420/20 nm bandpass filter to detect SHG signals from type I collagen in the specimens. Fresh femur specimens and composites were immersed in normal saline for inspection using a 20× water immersion lens. Figure 9a shows the normal rat femur bone structure under the two-photon microscope. The green woven fibers present well-aligned type I collagen fibers of the mature bone matrix, and the dark spindle-shaped areas might indicate the location of osteocytes. Figure 9b shows the SHG signals of the composite before implantation. The image was totally dark because there was no collagen inside. Figure 9c shows the SHG signals of the drill hole site which was implanted with the Gel-HA/HAP composite without SDF-1 for one month. Compared with the normal mature bone structure, the type I collagen fibers were looser and not so well aligned, which indicated the evidence of new bone formation inside the created bone defect. The dark areas in this picture might be the newly migrated osteocytes or residual HAP particles. Figure 9d shows the SHG signals of the drill hole site which was implanted with the Gel-HA/HAP composite with 100 ng/mL SDF-1. The signals of collagen fibers in composite with the SDF-1 group seem to be stronger then signals in composite without the SDF-1 group. Many various sized dark areas can also be seen between collagen fibers.

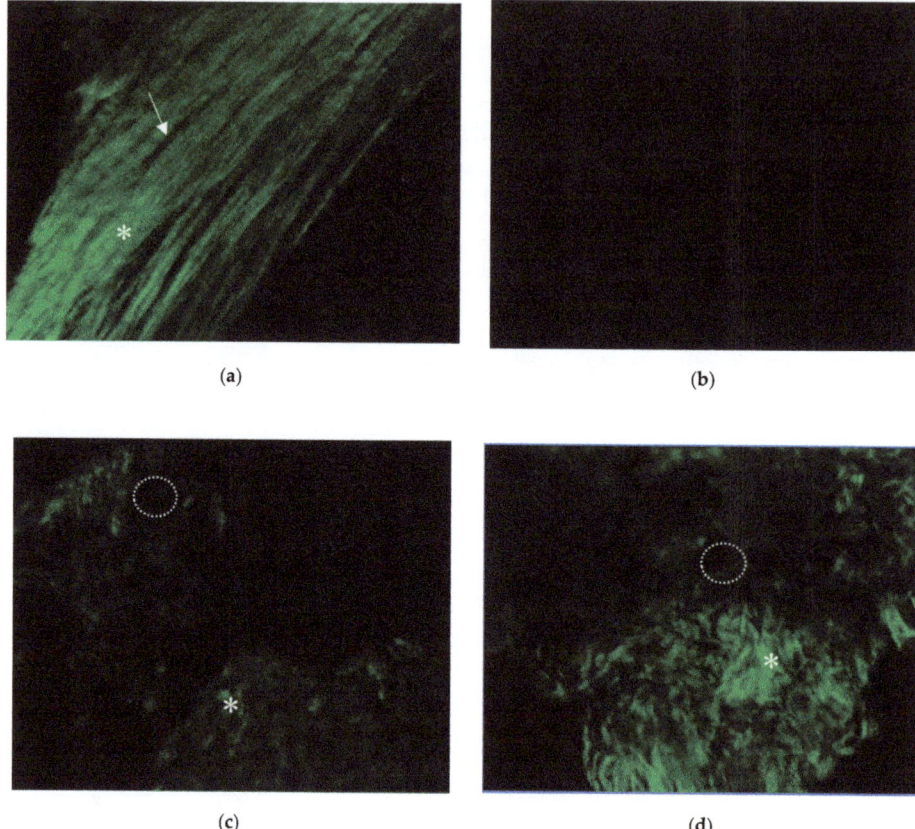

Figure 9. Two-photon excitation microscopy. (**a**) Second harmonic generation (SHG) signals of normal rat femur bone tissue. Green woven fibers (white asterisk) indicate well-aligned type I collagen fibers of mature bone matrix. Dark spindle-shaped areas (white arrow) in this image might present the location of osteocytes. (**b**) SHG signals of Gel-HA/CS/HAP composite before implantation. (**c**) SHG signals of the drill hole site filled with the composite without SDF-1. Some woven type I collagen fibers (asterisk) can be found, which indicate the evidence of new bone formation around the bone defect site. The dark areas (dotted circle) in this picture might be those osteocytes that just migrated in or the residual CS/HAP particles. (**d**) SHG signals of bone defect site implanted with the composite with 100 ng/mL SDF-1. Stronger signals of collagen fibers can be seen.

In this study, SHG imaging was used to confirm the bone ingrowth of the defect site filled with composites in an animal model. After a literature review, we found only a few articles about using two-photon excitation microscopy or SHG imaging on bone structure [35–37]. Due to the limited penetration depths of the laser beam, the complete bone structure cannot easily be visualized. Ishii et al. demonstrated dynamic live imaging of animal bone tissue under two-photon microscopy after injection of various fluorescent agents [35]. According to their study, both static histological information and the dynamic behaviors of live bone cells can be recorded under two-photon microscopy. In our study, however, no fluorescent agent was needed. Thanks to the SHG signals from type I collagen fibers, we could still successfully observe the normal bone structure of fresh rat femurs and the new bone growth on implanted composites. Using this technique, live SHG imaging of bone growth under two-photon microscopy may be possible in the future.

3.8.3. Micro-CT

The area of the bone defect site of each specimen was labeled under micro-CT sections (Figure 10). The percentages of new bone formation in each defect area were analyzed by CTan software and presented as BV/TV ratios (Figure 11). As we can see, although the bone had the ability of self-healing, regeneration in the control group was limited. Under the micro-CT section, the defect sites in the control group were still quite obvious after both one and two months. The BV/TV ratios of the control group after one and two months were 37.07% and 5.65%, respectively. For the micro-CT sections in the without SDF-1 group, some residual materials of the composites could still be seen, and there was more new bone formation in the defect area compared with the control group. The BV/TV ratios of the without SDF-1 group after one and two months were 75.06% and 41.89%, respectively. As for the 100 ng/mL SDF-1 group, its micro-CT sections revealed more bone formation compared with the other two groups. The BV/TV ratios of the percentages of the 100 ng/mL SDF-1 group after one and two months were 74.05% and 54.52%, respectively.

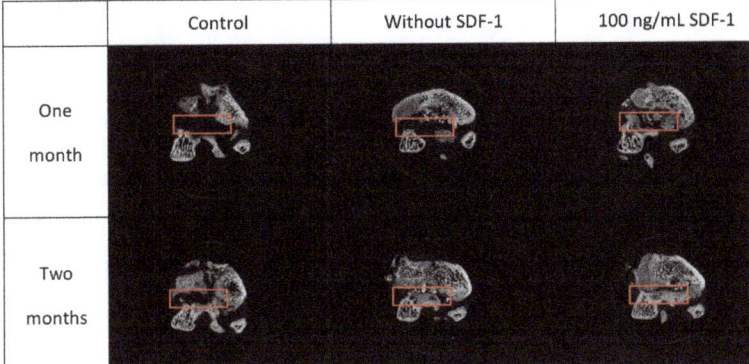

Figure 10. Micro-CT sections of specimens. Red rectangles indicate the site of the bone defect created by the drill. More bone ingrowth can be found in composite groups, especially the group with SDF-1.

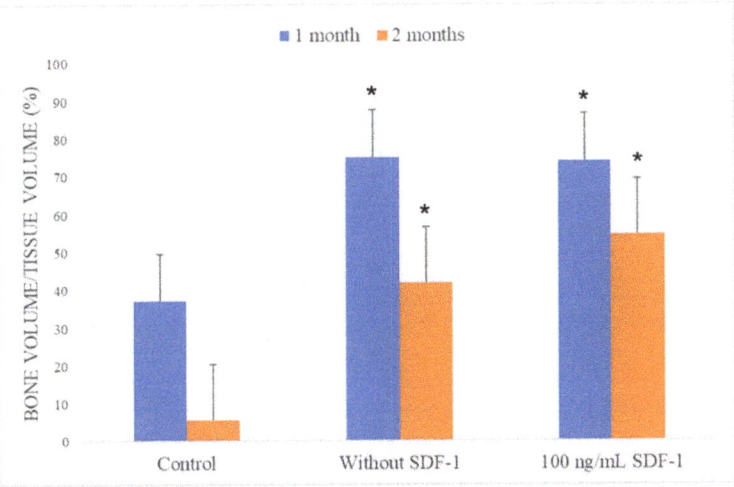

Figure 11. Bone volume/tissue volume (BV/TV) analysis of the scanned samples ($n = 3$). An obvious increase of new bone formation was noted in the composite without and composite with 100 ng/mL SDF-1 groups compared with the control group. * $p < 0.05$, compared with control group.

As we can see, the composites in both the without SDF-1 and 100 ng/mL SDF-1 groups were absorbed gradually. The results of the swelling and degradation tests suggest that after implantation, swelling resulted in material degradation, leading to absorption by cells. In addition, as it was shown in our micro-CT results, both composite groups without SDF-1 and with 100 ng/mL SDF-1 significantly had better bone healing compared with the control group. Compared with the composite without SDF-1 group after two months, the 100 ng/mL SDF-1 group showed a trend of better bone healing, but it did not achieve significance statistically. The probable reasons why it failed to show a significant difference between two kinds of composite include the limited number of experimental animals and suboptimal dose of SDF-1. In summary, the results of micro-CT analysis showed that Gel-HA/CS/HAP composites can enhance bone regeneration in an animal bone defect model. These results also implied that the MSC migration promoted by SDF-1 may be beneficial for bone healing, which still requires more studies to prove.

3.8.4. Histological Analysis

For histological analysis, the H&E and MT stain of the control group showed obvious bone defects even after two months (Figure 12a,b). Some fibrotic tissue ingrowth was also noted, and no new bone formation was found around the defect sites under MT stain. Histology sections of groups adding Gel-HA/CS/HAP composites with or without SDF-1 showed new bone formation under MT stain (Figure 12d,f). Residual materials were noted in both composite groups. Under H&E stain of both with and without SDF-1 groups, the residual materials were well integrated with normal bone tissue under MT stain (Figure 12c,e). Furthermore, the size of the residual material seemed to be smaller in the group with 100 ng/mL SDF-1 compared with the group without SDF-1.

In a study by Faruq et al. [38], biphasic calcium phosphate (CP) granules were combined with HA-Gel in a rabbit model, and promising bone regeneration was demonstrated by their CP/HA-Gel compounds in rabbit femur bone defects. An HAP/CS/HA composite encapsulated with collagenase (Col) was introduced by Subramaniam et al. for rat alveolar bone defects [39]. In their study, this HAP/CS/HA-Col composite improved new bone formation in a rat alveolar model. The use of a Gel-modified CS-HAP bone void filler with the addition of BMPs and bisphosphonate in a rat tibia bone defect model was reported by Teotia et al. [40]. Gelatin cement resulted in better cell proliferation in their study compared with cement without gelatin. In their rat alveolar defect model, maximum bone formation was found in animals implanted with cement incorporated with zoledronic acid. In our study, we integrated particles of calcium sulfate and hydroxyapatite directly into gelatin and hyaluronic acid as a composite. A rat femur bone defect model created by a trephine drill was used for our animal study. Comparable results were also achieved in our study, despite some differences in the study designs. Furthermore, there was no obvious systemic toxicity noted from our blood test results which could have been caused by the Gel-HA/CS/HAP composites with or without SDF-1. All the results showed that the use of Gel-HA/CS/HAP composites in bone defects seems to be safe and feasible, at least in a rat animal model.

The exact role of SDF-1 in bone formation remains unclear. As previously mentioned, in several studies, SDF-1 was found to have the ability to recruit MSCs [17–19]. Also, previous reports have shown the potential of MSCs in bone regeneration enhancement [17,20,21]. Even though the mechanism of MSCs during bone regeneration is not yet fully understood, we thought that the recruitment of MSCs may provide a certain degree of osteoinductive effect in the bone healing process—something which may require further investigation to confirm [41]. Shen et al. used a silk fibroin–nanohydroxyapatite scaffold with sustained release of SDF-1 and BMP-2 on a rat cranial bone defect model [42]. They successfully proved that combining SDF-1 and BMP-2 results in a synergistic effect on bone regeneration. In our study, cbMSC-hTERT cells also obviously migrated toward SDF-1-releasing material compared with the control groups. Furthermore, adding SDF-1 into the Gel-HA/CS/HAP composite showed comparable results to the pure composite group and even revealed a trend toward better bone regeneration.

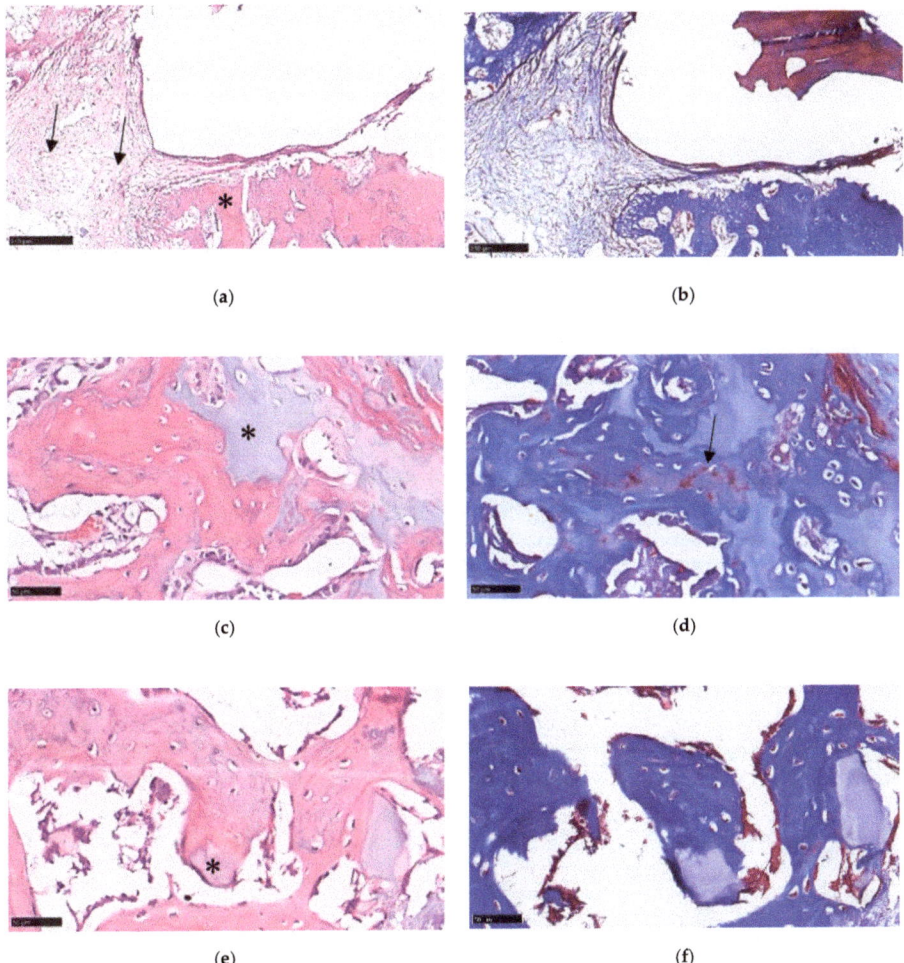

Figure 12. (a) Histology section under 10× magnification view showed a large bone defect and normal bone structure (asterisk) in the control group under hematoxylin and eosin (H&E) staining. Some fibrotic tissue (black arrows) was also found in the bone defect. (b) No new bone formation was noted under Masson's trichrome (MT) stain in the control group. (c) Under 40× magnification view of the composite without SDF-1 group, residual material (asterisk) can be seen. (d) New bone formation (black arrow) can be found under MT stain in the group without SDF-1. (e) In the 100 ng/mL SDF-1 group, smaller residual material (asterisk) was shown to be well integrated into the normal bone structure under H&E stain. (f) Histology section of SDF-1 group under MT stain.

The size, hardness, and applicability of bone void fillers are quite important when dealing with bone defects in clinical practice [3,4]. It is important that the materials be easy to manipulate and handle during the implantation of bone fillers into bone defect sites. When the material is too large, it may require more time to fabricate or trim it to the correct size before implantation. Also, it may be difficult to implant into the bone defect if the material is too hard. However, if the material is too small or too soft, the material may be easily spread out around the operation field during implantation. In that situation, heterotrophic bone formation or soft tissue irritation may develop. In this study, the Gel-HA/CS/HAP composite had adequate but not excessive hardness and could be easily cut into a suitable size without difficulty. The composite could be handled and implanted into the bone defect site without it spreading out. Nevertheless, the composite could be retained inside the bone defect site without dissolution or dislocation, even after saline or blood flushes, which is quite important during real operations on humans. The good applicability and sustainability of this composite appears to make it suitable for clinical use.

There are some limitations to our study. Statistical analysis was difficult because of the small number of animal models. Furthermore, due to the fast bone generation speed of rats, shorter observation intervals may be needed for more detailed comparison. In this study, human SDF-1 was used on the rat model for the in vivo test. According to a previous study, human and murine SDF-1 are highly similar and can react across species [43]. However, the MSC recruitment ability of human SDF-1 may still possibly be compromised in a rat model.

4. Conclusion

In summary, the good biocompatibility and biodegradability of Gel-HA/CS/HAP composites with or without SDF-1 were both confirmed by in vitro tests in our study. Enhanced bone regeneration after implantation of Gel-HA/CS/HAP composites was also observed in micro-CT and histology analyses in our in vivo study. The composites with SDF-1 showed a trend of better bone healing compared with the composite without SDF-1, but the difference was not statistically significant. In conclusion, this Gel-HA/CS/HAP composite with SDF-1 may be a safe and feasible material to use as a bone void filler in bone defects. Further studies are still needed to reveal the definite pathway by which SDF-1 enhances bone growth and the best dose of SDF-1 which should be added to the composite.

Author Contributions: Conceptualization and supervision, F.-H.L.; investigation and writing—original draft preparation, C.-Y.H.; resources and software, C.-Y.Y.; funding acquisition, project administration, and writing—review and editing, Y.-L.C.

Funding: This research was supported by the National Taiwan University Hospital (grant no. 107-004004). The sponsor had no role in the study design, data collection, data analysis, data interpretation, or manuscript writing.

Conflicts of Interest: The authors declare no conflict of interest.

References

1. Nauth, A.; McKee, M.D.; Einhorn, T.A.; Watson, J.T.; Li, R.; Schemitsch, E.H. Managing bone defects. *J. Orthop. Trauma* **2011**, *25*, 462–466. [CrossRef]
2. Fillingham, Y.; Jacobs, J. Bone grafts and their substitutes. *Bone Jt. J.* **2016**, *98-B*, 6–9. [CrossRef]
3. Lobb, D.C.; DeGeorge, B.R., Jr.; Chhabra, A.B. Bone graft substitutes: Current concepts and future expectations. *J. Hand Surg. Am.* **2019**, *44*, 497–505.e492. [CrossRef]
4. Wang, W.; Yeung, K.W.K. Bone grafts and biomaterials substitutes for bone defect repair: A review. *Bioact. Mater.* **2017**, *2*, 224–247. [CrossRef]
5. Dressman, H. Uber knochenplombierung. *Beitr Klin Chir* **1892**, *9*, 804–810.
6. Kelly, C.M.; Wilkins, R.M.; Gitelis, S.; Hartjen, C.; Watson, J.T.; Kim, P.T. The use of a surgical grade calcium sulfate as a bone graft substitute: Results of a multicenter trial. *Clin. Orthop. Relat. Res.* **2001**, *382*, 42–50. [CrossRef]
7. Pietrzak, W.S.; Ronk, R. Calcium sulfate bone void filler: A review and a look ahead. *J. Craniofac. Surg.* **2000**, *11*, 327–333. [CrossRef]
8. Jepegnanam, T.S.; von Schroeder, H.P. Rapid resorption of calcium sulfate and hardware failure following corrective radius osteotomy: 2 case reports. *J. Hand Surg. Am.* **2012**, *37*, 477–480. [CrossRef]
9. Lee, G.H.; Khoury, J.G.; Bell, J.E.; Buckwalter, J.A. Adverse reactions to osteoset bone graft substitute, the incidence in a consecutive series. *Iowa Orthop. J.* **2002**, *22*, 35–38.
10. Hing, K.A.; Wilson, L.F.; Buckland, T. Comparative performance of three ceramic bone graft substitutes. *Spine J.* **2007**, *7*, 475–490. [CrossRef]
11. Holmes, R.E. Bone regeneration within a coralline hydroxyapatite implant. *Plast. Reconstr. Surg.* **1979**, *63*, 626–633. [CrossRef]

12. Even, J.; Eskander, M.; Kang, J. Bone morphogenetic protein in spine surgery: Current and future uses. *J. Am. Acad. Orthop. Surg.* **2012**, *20*, 547–552.
13. Gruskin, E.; Doll, B.A.; Futrell, F.W.; Schmitz, J.P.; Hollinger, J.O. Demineralized bone matrix in bone repair: History and use. *Adv. Drug Deliv. Rev.* **2012**, *64*, 1063–1077. [CrossRef]
14. Shehadi, J.A.; Elzein, S.M. Review of commercially available demineralized bone matrix products for spinal fusions: A selection paradigm. *Surg. Neurol. Int.* **2017**, *8*, 203. [CrossRef]
15. Boyce, T.; Edwards, J.; Scarborough, N. Allograft bone. The influence of processing on safety and performance. *Orthop. Clin. N. Am.* **1999**, *30*, 571–581. [CrossRef]
16. Finkemeier, C.G. Bone-grafting and bone-graft substitutes. *J. Bone Jt. Surg. Am.* **2002**, *84*, 454–464. [CrossRef]
17. Kitaori, T.; Ito, H.; Schwarz, E.M.; Tsutsumi, R.; Yoshitomi, H.; Oishi, S.; Nakano, M.; Fujii, N.; Nagasawa, T.; Nakamura, T. Stromal cell-derived factor 1/cxcr4 signaling is critical for the recruitment of mesenchymal stem cells to the fracture site during skeletal repair in a mouse model. *Arthritis Rheum. Off. J. Am. Coll. Rheumatol.* **2009**, *60*, 813–823. [CrossRef]
18. Otsuru, S.; Tamai, K.; Yamazaki, T.; Yoshikawa, H.; Kaneda, Y. Circulating bone marrow-derived osteoblast progenitor cells are recruited to the bone-forming site by the cxcr4/stromal cell-derived factor-1 pathway. *Stem Cells* **2008**, *26*, 223–234. [CrossRef]
19. Segers, V.F.M.; Tokunou, T.; Higgins, L.J.; MacGillivray, C.; Gannon, J.; Lee, R.T. Local delivery of protease-resistant stromal cell derived factor-1 for stem cell recruitment after myocardial infarction. *Circulation* **2007**, *116*, 1683–1692. [CrossRef]
20. Knight, M.N.; Hankenson, K.D. Mesenchymal stem cells in bone regeneration. *Adv. Wound Care* **2013**, *2*, 306–316. [CrossRef]
21. Jin, Y.Z.; Lee, J.H. Mesenchymal stem cell therapy for bone regeneration. *Clin. Orthop. Surg.* **2018**, *10*, 271–278. [CrossRef]
22. Kang, H.W.; Tabata, Y.; Ikada, Y. Fabrication of porous gelatin scaffolds for tissue engineering. *Biomaterials* **1999**, *20*, 1339–1344. [CrossRef]
23. Chen, W.Y.J.; Abatangelo, G. Functions of hyaluronan in wound repair. *Wound Repair Regen.* **1999**, *7*, 79–89. [CrossRef]
24. Akao, M.; Aoki, H.; Kato, K. Mechanical-properties of sintered hydroxyapatite for prosthetic applications. *J. Mater. Sci.* **1981**, *16*, 809–812. [CrossRef]
25. Shyong, Y.J.; Wang, M.H.; Kuo, L.W.; Su, C.F.; Kuo, W.T.; Chang, K.C.; Lin, F.H. Mesoporous hydroxyapatite as a carrier of olanzapine for long-acting antidepression treatment in rats with induced depression. *J. Control. Release* **2017**, *255*, 62–72. [CrossRef]
26. Hung, C.J.; Yao, C.L.; Cheng, F.C.; Wu, M.L.; Wang, T.H.; Hwang, S.M. Establishment of immortalized mesenchymal stromal cells with red fluorescence protein expression for in vivo transplantation and tracing in the rat model with traumatic brain injury. *Cytotherapy* **2010**, *12*, 455–465. [CrossRef]
27. Liu, M.; Lv, Y. Reconstructing bone with natural bone graft: A review of in vivo studies in bone defect animal model. *Nanomaterials* **2018**, *8*, 999. [CrossRef]
28. Hollinger, J.O.; Kleinschmidt, J.C. The critical size defect as an experimental model to test bone repair materials. *J. Craniofac. Surg.* **1990**, *1*, 60–68. [CrossRef]
29. Kondo, N.; Ogose, A.; Tokunaga, K.; Ito, T.; Arai, K.; Kudo, N.; Inoue, H.; Irie, H.; Endo, N. Bone formation and resorption of highly purified beta-tricalcium phosphate in the rat femoral condyle. *Biomaterials* **2005**, *26*, 5600–5608. [CrossRef]
30. Li, Y.; Chen, S.-K.; Li, L.; Qin, L.; Wang, X.-L.; Lai, Y.-X. Bone defect animal models for testing efficacy of bone substitute biomaterials. *J. Orthop. Transl.* **2015**, *3*, 95–104. [CrossRef]
31. Schmitz, J.P.; Hollinger, J.O. The critical size defect as an experimental-model for craniomandibulofacial nonunions. *Clin. Orthop. Relat. R.* **1986**, *205*, 299–308. [CrossRef]
32. Sun, J.; Mou, C.; Shi, Q.; Chen, B.; Hou, X.; Zhang, W.; Li, X.; Zhuang, Y.; Shi, J.; Chen, Y.; et al. Controlled release of collagen-binding sdf-1alpha from the collagen scaffold promoted tendon regeneration in a rat achilles tendon defect model. *Biomaterials* **2018**, *162*, 22–33. [CrossRef]
33. Sun, J.; Zhao, Y.; Li, Q.; Chen, B.; Hou, X.; Xiao, Z.; Dai, J. Controlled release of collagen-binding sdf-1alpha improves cardiac function after myocardial infarction by recruiting endogenous stem cells. *Sci. Rep.* **2016**, *6*, 26683. [CrossRef]

34. Wang, W.; Li, W.; Ong, L.L.; Furlani, D.; Kaminski, A.; Liebold, A.; Lutzow, K.; Lendlein, A.; Wang, J.; Li, R.K.; et al. Localized sdf-1alpha gene release mediated by collagen substrate induces cd117 stem cells homing. *J. Cell. Mol. Med.* **2010**, *14*, 392–402. [CrossRef]
35. Ishii, M.; Fujimori, S.; Kaneko, T.; Kikuta, J. Dynamic live imaging of bone: Opening a new era with 'bone histodynametry'. *J. Bone Miner. Metab.* **2013**, *31*, 507–511. [CrossRef]
36. Sano, H.; Kikuta, J.; Furuya, M.; Kondo, N.; Endo, N.; Ishii, M. Intravital bone imaging by two-photon excitation microscopy to identify osteocytic osteolysis in vivo. *Bone* **2015**, *74*, 134–139. [CrossRef]
37. Campagnola, P.J.; Loew, L.M. Second-harmonic imaging microscopy for visualizing biomolecular arrays in cells, tissues and organisms. *Nat. Biotechnol.* **2003**, *21*, 1356–1360. [CrossRef]
38. Faruq, O.; Kim, B.; Padalhin, A.R.; Lee, G.H.; Lee, B.T. A hybrid composite system of biphasic calcium phosphate granules loaded with hyaluronic acid-gelatin hydrogel for bone regeneration. *J. Biomater. Appl.* **2017**, *32*, 433–445. [CrossRef]
39. Subramaniam, S.; Fang, Y.-H.; Sivasubramanian, S.; Lin, F.-H.; Lin, C.-P. Hydroxyapatite-calcium sulfate-hyaluronic acid composite encapsulated with collagenase as bone substitute for alveolar bone regeneration. *Biomaterials* **2016**, *74*, 99–108. [CrossRef]
40. Teotia, A.K.; Gupta, A.; Raina, D.B.; Lidgren, L.; Kumar, A. Gelatin-modified bone substitute with bioactive molecules enhance cellular interactions and bone regeneration. *ACS Appl. Mater. Interfaces* **2016**, *8*, 10775–10787. [CrossRef]
41. Lin, W.P.; Xu, L.L.; Zwingenberger, S.; Gibon, E.; Goodman, S.B.; Li, G. Mesenchymal stem cells homing to improve bone healing. *J. Orthop. Transl.* **2017**, *9*, 19–27. [CrossRef]
42. Shen, X.; Zhang, Y.; Gu, Y.; Xu, Y.; Liu, Y.; Li, B.; Chen, L. Sequential and sustained release of sdf-1 and bmp-2 from silk fibroin-nanohydroxyapatite scaffold for the enhancement of bone regeneration. *Biomaterials* **2016**, *106*, 205–216. [CrossRef]
43. Lapidot, T. Mechanism of human stem cell migration and repopulation of nod/scid and b2mnull nod/scid mice. The role of sdf-1/cxcr4 interactions. *Ann. N. Y. Acad. Sci.* **2001**, *938*, 83–95. [CrossRef]

© 2019 by the authors. Licensee MDPI, Basel, Switzerland. This article is an open access article distributed under the terms and conditions of the Creative Commons Attribution (CC BY) license (http://creativecommons.org/licenses/by/4.0/).

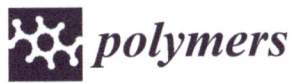

Article

Gamma-Irradiation-Prepared Low Molecular Weight Hyaluronic Acid Promotes Skin Wound Healing

Yu-Chih Huang [1], Kuen-Yu Huang [1], Wei-Zhen Lew [1], Kang-Hsin Fan [2], Wei-Jen Chang [1] and Haw-Ming Huang [1,3,4,*]

1. School of Dentistry, College of Oral Medicine, Taipei Medical University, Taipei 11031, Taiwan
2. Dental Department, En Chu Kong Hospital, New Taipei City 23741, Taiwan
3. Graduate Institute of Biomedical Optomechatronics, College of Biomedical Engineering, Taipei 11031, Taiwan
4. Research Center of Biomedical Device, Medical University, Taipei 11031, Taiwan
* Correspondence: hhm@tmu.edu.tw; Tel.: +886-291-937-9783

Received: 4 July 2019; Accepted: 17 July 2019; Published: 19 July 2019

Abstract: In this study, we prepared low-molecular-weight hyaluronic acid (LMWHA) powder by γ-irradiation. The chemical and physical properties of γ-irradiated LMWHA and the in vitro cellular growth experiments with γ-irradiated LMWHA were analyzed. Then, hyaluronic acid exposed to 20 kGy of γ-irradiation was used to fabricate a carboxymethyl cellulose (CMC)/LMWHA fabric for wound dressing. Our results showed that γ-irradiated LMWHA demonstrated a significant alteration in carbon–oxygen double bonding and can be detected using nuclear magnetic resonance and ultraviolet (UV)-visible (Vis) spectra. The γ-irradiated LMWHA exhibited strain rate-dependent Newton/non-Newton fluid biphasic viscosity. The viability of L929 skin fibroblasts improved upon co-culture with γ-irradiated LMWHA. In the in vivo animal experiments, skin wounds covered with dressings prepared by γ-irradiation revealed acceleration of wound healing after two days of healing. The results suggest that γ-irradiated LMWHA could be a potential source for the promotion of skin wound healing.

Keywords: low molecular weight; hyaluronic acid; wound healing; gamma ray; membrane

1. Introduction

Wound healing is a series of processes that involves the control of inflammation, cell migration and new tissue remodeling [1,2]. It is reported that a material with anti-inflammatory, antimicrobial and antioxidant properties that promotes cell migration can serve as a potential solution for treating skin and soft-tissue wounds [3].

Hyaluronic acid (HA) is a biopolymer found mainly in the extracellular space [4] and joints [5,6]. The primary physiological function of HA is its buffering action, which is due to its excellent viscoelastic properties after water absorption [7]. Thus, traditionally, HA was reportedly used as a medical material for retaining skin moisture and for osteoarthritis therapy [8]. Recently, studies have shown that HA exhibits anti-inflammatory and antibacterial activities [9,10]. In addition, since free radicals can break down hyaluronic acid into smaller fragments in damaged tissues, it also has the antioxidant function of scavenging free radicals. Furthermore, it is well known that HA also can be a scaffold during tissue repair to provide cell climbing and migration opportunities [11]. With these useful functions, HA is reported to accelerate the process of wound healing [10,12,13].

Recently, many studies have investigated the association between the molecular weight of HA and its physiological functions [14,15]. In the initial stage of wound healing, high-molecular-weight HA (HMWHA) (~2000 kDa) accumulates in the extracellular matrix and binds to fibrinogen to form a clot. Thereafter, in the inflammatory stage, HMWHA is broken down into low-molecular-weight hyaluronic acid (LMWHA) (80–800 kDa) by hyaluronidase for subsequent use in healing [16–18]. At this stage, LMWHA is reported to participate in the inflammatory response, involving macrophage activation and chemokine expression [11]. D'Agostino et al. (2015) performed an in vitro study and concluded that LMWHA accelerated wound repair because it inhibited fibroblast differentiation and collagen deposition at this early stage [11]. These effects allow macrophages to move to the wound site to phagocytose debris and clean infectious matter [16]. Several investigations also found that LMWHA prevented oxygen free radical damage to granulation tissue [19] and increased the self-defense of skin epithelium by inducing various skin-repair-related genes [20] during the wound healing process.

It is proposed that although HMWHA is used in various medical sciences, LMWHA may provide potential beneficial effects for wound healing. However, until now, the preparation of HA with a specific molecular weight has been a complex work that is not easy to control [21]. To fabricate LMWHA efficiently, several scholars used physical (ultrasound, ozone, electron beam, γ-irradiation and thermal treatment) and chemical methods (enzyme and acid degradation) to break the primary bond of HMWHA [21–23]. Among these methods, γ-irradiation is reported to reduce the molecular weight of HA without structural alteration of the polymer [21,23–25].

It was reported that, even though the main structure of HA fragments remained intact, the water-absorbing ability was changed due to the molecular weight reduction of the polymer [23,24]. However, the in vivo evidence supporting the efficacy of using LMWHA to fabricate a wound dressing membrane is still limited. Accordingly, the purpose of this study was to prepare LMWHA powder by γ-irradiation. The prepared LMWHA was used as a material to fabricate a hydrogel dressing membrane. We hypothesized that the γ-irradiated powder could be a useful material for fabricating a hydrogel membrane for skin wound dressing.

2. Materials and Methods

2.1. Preparation of Low-Molecular-Weight Hyaluronic Acid

The HA used in this study (molecular weight (MW) 3000 kDa) was purchased from Cheng-Yi Chemical Industry Co. Ltd. (Taipei, Taiwan). Before the experiments, the HA powder, stored in tightly capped tubes, was irradiated using a cobalt-60 irradiator (Point Source, AECL, IR-79, Nordion, Canada) at 22 °C, with a dose rate of 1 kGy/h at the sample location. The γ-irradiated HA powder was divided into four groups. The first two HA groups were γ-irradiated with a dose of 20 kGy (20 h exposure) once (HA20I) and twice (HA20II), respectively. The third and fourth LMWHA groups were exposed to γ-irradiation at doses of 40 kGy (HA40) and 60 kGy (HA60) for 20 h, respectively. The unexposed HMWHA powder served as the control group (HA0). The irradiation dose was confirmed using alanine pellet dosimeters (FWT-50, Far West Technology, Inc., Goleta, CA, USA).

The molecular weights of the HA with and without γ-irradiation were measured by gel permeation chromatography. In this study, each group of HA powder was formulated into a 10 mg/mL HA solution in 0.1 M NaCl, and then 200 µL of the sample was injected into a separation module (Series 200, Perkin Elmer, Waltham, MA, USA) equipped with a chromatography column (SB-806M HQ, Shodex, Kanagawa, Japan). A refractive index (RI) detector (Series 200, Perkin Elmer, Waltham, MA, USA) was used to detect the signals. The mobile phase was 0.1 M sodium nitrate (purity: 99.9%, Merck KGaA, Darmstadt, Germany). The flow rate was 0.5 mL/min, and the analyses were performed at 25 °C. The calibration was achieved using a standard kit (Pullulan ReadyCal Kits, PSS Polymer Standards Service, Mainz, Germany). The gel permeation chromatography data were collected and analyzed using commercially available software (ChromManager 5.8, ABDC WorkShop, Taichung, Taiwan). The dispersity of each sample was obtained by calculating the ratio between weight average molecular weight (M_w) and the number average molecular weight (M_n) for different samples.

2.2. Chemical Property Analysis

The ^{13}C nuclear magnetic resonance (NMR) spectra of the γ-ray treated and untreated HA were recorded at 27 °C on a 500 MHz NMR spectrometer (DRX500 Avance, Bruker BioSpin GmbH, Rheinstetten, Germany). D2O (Sigma-Aldrich, St. Louis, MO, USA) was used as the solvent in all the NMR experiments. Fourier-transform infrared (FT-IR) spectra of the samples were detected using an infrared spectrophotometer (Spectrum one, Perkin Elmer, Waltham, MA, USA). Before tests, the γ-ray treated and untreated HA powders were mixed with KBr (Sigma-Aldrich, St. Louis, MI, USA) and compressed into disks. The wavelength range was set at 650–4000 cm^{-1}. Transmission mode spectra were obtained from 24 scans. To detect the UV-Vis absorption spectra of the γ-irradiated HA, samples were diluted in distilled water to a concentration of 0.2% (mg/mL). UV-Vis spectra were measured using a CT-2400 Spectrophotometer (Great Tide Instrument Co., Ltd., Taipei, Taiwan) at a wavelength range of 200 nm to 500 nm. During detection, distilled water was used as a reference.

2.3. Physical Property Detection

To determine the pH of the variously irradiated HA samples, the samples were diluted 1:500 in purified water and stirred for 12 h. The pH of the samples was measured with a pH meter at room temperature (Model 6173, JENCO Quality Instruments, San Diego, CA, USA) equipped with a pH electrode (HI1413, Hanna Instruments, Inc., Woonsocket, RI, USA). Before the tests, the pH meter was calibrated with pH 7 and pH 4 buffers. The rheological characteristic of the tested HA samples (prepared to 2% mg/mL solution) was measured using an oscillatory rheometer at 25 °C (Rheostress 1, Haake, Karlsruhe, Germany). The frequency range and shear ratio were set at 0.1–100 Hz. The dynamic viscosity (η^*) of the tested HA was recorded as a function of the strain rate.

The moisture absorption test method was modified from that of a previous study [7]. The irradiated HA powders were dried in an oven for 24 h before testing the moisture absorption properties. Then, 0.1 g of dried HA sample was put in a 3.5 cm culture plate and placed in an incubator (REVCO RCO3000T, Thermo Fisher Scientific, Waltham, MA, USA) at a temperature of 37 °C and relative humidity of 95%. Samples were weighted every 24 h. The water absorption capacity was expressed by the change in weight of the material after the moisture had been absorbed.

2.4. In Vitro Cell Viability Experiments

A cell viability assay was performed to test the effect of the γ-irradiated HA samples on the viability of skin cells. The skin fibroblast cell line L929 (American Type Culture Collection, ATCC, no. CCL-1) was used for this in vitro cell analysis. The cells were seeded in 24-well plates at a concentration of 2×10^4 cells/mL and were maintained in Dulbecco's modified Eagle medium supplemented with L-glutamine and 10% fetal bovine serum (DMEM, Gibco, Grand Island, NY, USA). The cells were cultured in an incubator in an environment of 5% CO2 at 37 °C and 100% humidity. The viability of the L929 cells co-cultured with 0.1% γ-irradiated HA for six days was detected using the tetrazolium salt method (MTT, Sigma-Aldrich, St. Louis, MO, USA). Briefly, after the test cells were incubated with tetrazolium salt for 4 h, 500 μL dimethyl sulfoxide (DMSO, Sigma-Aldrich) was added and incubated overnight to solubilize the formazan dye. The optical density was determined using a microplate reader (EZ Read 400, Biochrom, Holliston, MA, USA) at a wavelength of 570/690 nm.

2.5. In Vivo Wound Healing Tests

2.5.1. Carboxymethyl Cellulose (CMC)/LMWHA Dressing Fabrication

A CMC/LMWHA hydrogel was prepared on a nonwoven fabric as described in a previous study [26] to test the wound healing effect of the prepared LMWHA. The 15 mL of CMC (10 mg/mL) and the 15 mL of LMWHA (30 mg/mL) were mixed with a magnetic stirrer. Then, the mixture was moved to the surface of a nonwoven fabric (3 cm × 3 cm). The dressings were then put into an oven at 37 °C for 3 h to form dried hydrogel dressings. In this study, due to the chemical and physical experiments and the policy to reduce animal use based on the Helsinki Declaration, only the HA20I sample was used to fabricate the dressing for use in the animal study. Dressings prepared with CMC only were used in the control group.

2.5.2. In Vivo Wound Healing Experiment

Eight healthy male Sprague Dawley rats weighing 210 to 290 g were used to assess the effects of LMWHA on wound healing. The rats were obtained from the Laboratory Animal Center at the National Applied Research Laboratories (Hsinchu, Taiwan). They were kept in hygienic cages and maintained with a 12 h light/dark cycle. The study protocol and procedure were reviewed and approved by the Institution Animal Care and Use Committee (IACUC Approval No. L10708), and all efforts were made to minimize the number of rats and suffering to produce reliable scientific data.

Before experimentation, the rats' backs were shaved (5 cm × 5 cm) with an electric animal shaver, and 75% alcohol was used to avoid infection. The rats were anesthetized with 5% isoflurane in an anesthesia induction chamber. One linear incision wound with an area of 2 cm × 2 cm was made on the shaved area using sterile scissors. The SD rats were randomly divided into two groups, with four rats each in experimental and control group. For the experimental group, the wound sites of the rats were covered with the dressing prepared with LMWHA (HA20I). The HA-free dressing was applied to the wounds of the control animals. The covered dressings were replaced every two days during the 12 day experimental period. The rats were housed individually and kept at an environmental temperature of 21 °C and a humidity of 60 to 70% during the entire experimental period. The wound of each rat was photographed every two days with a digital camera. The recorded wound areas were measured using ImageJ software (National Institutes of Health, Bethesda, Rockville, MD, USA). The wound size was expressed as a percentage reduction of the original wound size.

2.6. Statistical Analysis

For cell viability and animal tests, mean values and standard deviations of each measurement were recorded. One-way analysis of variance (ANOVA) with Tukey's post hoc and Student t-tests (SPSS Inc., Chicago, IL, USA) were performed to evaluate the changes between the samples and controls, for cell and animal experiments, respectively. A p-value lower than 0.05 was considered statistically significant.

3. Results and Discussion

As shown in Table 1, the molecular weights of the HA samples exposed to γ-irradiation decreased significantly in a dose-dependent manner. The molecular weight was 232.4 kDa when 20 kGy γ-radiation was applied. This value decreased to 141.8 kDa and 59.5 kDa when 40 kGy and 60 kGy γ-radiation were used, respectively. This phenomenon was similar to the results of a previous HA powder experiment [23]. However, our data were much lower than the findings of Kim et al. (2008), who used 50 kGy γ-irradiated HA dissolved in distilled water, for which the molecular weight decreased to 6.5 kDa [24]. This extreme decrease is because hydrogen and hydroxide radicals formed during the irradiation of the water, breaking the molecular chain of the HA molecules [24]. As mentioned above, there are several methods to reduce the molecular weight of HA. Among these methods, enzymatic and chemical methods are relatively uncontrollable. LMWHA prepared by these means is reported to show a broader molecular weight distribution compared to that prepared by physical techniques (Kim et al., 2008). For definition, the HA20II and HA40 samples received the same dose of γ-radiation. However, from Table 1, we found that the molecular weight and pH values of HA20II were lower than those of HA40. This may be due to the fact that the position and exposed direction of the HA20II sample was changed at the time interval between the two exposures. This procedure makes the samples received a more homogeneous γ-radiation and leads to a more serious breakdown of their molecular chains. In the present study, the polydispersity (M_w/M_n) decreased along with the γ-ray dose. This value reduced from 386.2 to 3.7 when 20 kGy γ-radiation was used. Since the polydispersity of a polymer is an important parameter related to degradation conditions and molecular weight distribution, our results confirm the conclusions of previous studies showing that LMWHA manufactured by γ-irradiation degrades the HMWHA powder more randomly [23,24] and makes the material more homogeneous when dissolved in water.

Table 1. Molecular weights, dispersity and pH values of LMWHA irradiated with various doses of γ-irradiation.

Sample	Molecular Weight (kDa)	Dispersity (Mw/Mn)	pH
HA0	2983.7	386.2	6.76
HA20I	232.4	3.7	5.95
HA20II	99.2	3.4	5.49
HA40	141.8	3.2	5.54
HA60	59.5	2.5	5.2

The samples were analyzed by FT-IR, ^{13}C NMR and UV-Vis spectroscopy to confirm the LMWHA structural changes due to γ-irradiation. For the ^{13}C NMR analysis, the major difference between the untreated HMWHA and γ-irradiated LMWHA can be found at chemical shifts of 171 and 175 ppm (Figure 1). According to previous reports, these peaks are due to carbon–oxygen double bonds (C=O). The 171 and 175 ppm were carboxylate carbon and acetamido carbonyl carbon, respectively [14,27,28]. When the HA samples were exposed to γ-rays, the peak ratio of 175/171 ppm markedly increased. The analysis of UV-Vis spectra (Figure 2) confirmed previous findings [23,24] that γ-irradiation increases the absorbance at 265 nm. A prior report on the effects of γ-irradiation on alginates using 60Co in the dosage range of 20 to 500 kGy indicated that the absorbance at 265 nm is due to the double bond of HA formed after the degradation of the main chain of the polymer. This effect may be attributed to a hydrogen abstraction reaction after degradation [29]. That is, the γ-irradiation of LMWHA significantly changes the chemical structure of the HA associated with carbon–oxygen double bonds. Interestingly, not all degradation methods have the same effect of increasing C=O bonding. For example, HA treated with ultrasound, hydrogen peroxide and ozone showed no apparent changes in NMR and UV-Vis spectra [21,25].

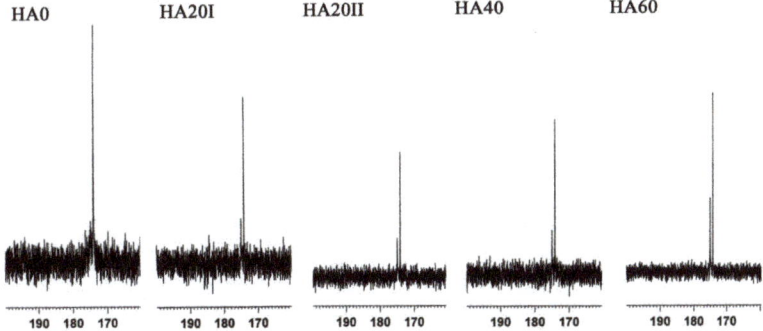

Figure 1. ^{13}C NMR spectra of γ-ray treated and untreated hyaluronic acid (HA) samples at chemical shifts between 160 and 190 ppm.

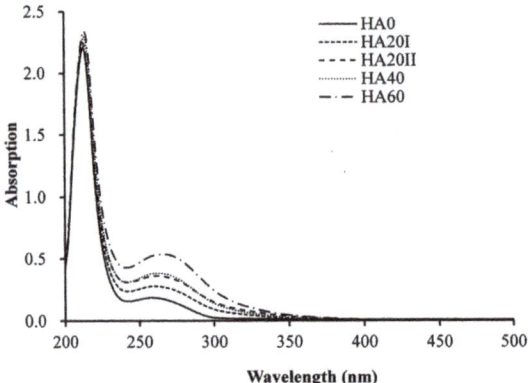

Figure 2. UV-Vis spectra of HA irradiated with various doses of γ-radiation. HA20I and HA20II are HA samples that received 20 kGy γ-irradiation once or twice, respectively. HA40 and HA60 represent the HA samples that received 40 kGy and 60 kGy γ-irradiation, respectively.

The FT-IR spectra of LMWHA degraded from various doses of γ-irradiation are shown in Figure 3. According to previous reports [23–25], the absorption bands at 1061–1166 cm^{-1} are characteristic f carbohydrates. The band at 1673 cm^{-1} is associated with carbon–oxygen double bonds (C=O). The bands at 1632 cm^{-1}, 1578 cm^{-1} and 1320 cm^{-1} correspond to amides. No substantial change was found when comparing the FT-IR spectra of the HMWHA (HA0) to the γ-irradiated LMWHA samples. This phenomenon differs from that of the 13C NMR and UV-Vis tests. This result may be due to the carbon–oxygen double bond-associated band (1673 cm^{-1}), which already exists in untreated HMWHA and can overlap with other bands, making it hard to distinguish the molecular size of the HA.

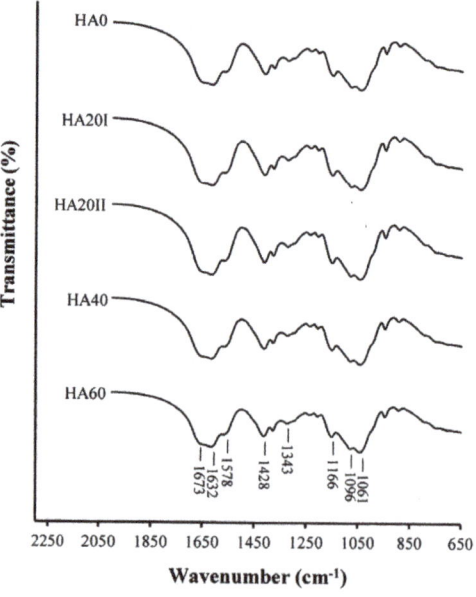

Figure 3. FTIR spectra of high-molecular-weight hyaluronic acid (HMWHA) and low-molecular-weight hyaluronic acid (LMWHA) treated with various doses of γ-ray irradiation.

We confirmed the conclusion of previous studies that LMWHA prepared by γ-irradiation preserves its fundamental structure, but with the formation of large amounts of the carbonyl group due to the depolymerization process [30]. This depolymerization of HA results in decreased pH values and the decreased viscosity of HA, as shown in Table 1 and Figure 4 [24,28]. In Figure 4, untreated HMWHA shows a typical viscosity pattern. The dynamic viscosity of HMWHA depends on the shear rate as a non-Newtonian liquid [31]. However, the dynamic viscosity of γ-irradiated HA markedly decreased independently with the applied shear rate when the rate was less than 10 s^{-1}. That is, at this status, the γ-irradiated HA showed a Newtonian liquid viscosity behavior. This effect can be attributed to the depolymerization process due to γ-irradiation, which results in the collapse of the macromolecular coils [31]. Interestingly, when the strain rate was larger than 10 s^{-1}, the γ-irradiated HA demonstrated a shear-thickening characteristic. The viscosity of the prepared LMWHA increased with increasing strain rate. This "strain-hardening" property suggests that the LMWHA fluids provide a shock-damping function and protective effect when a sudden high-load impact is applied [32].

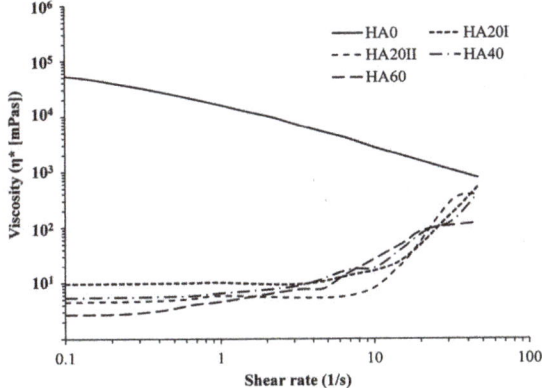

Figure 4. Dynamic viscosity dependencies of applied shear rates of the HMWHA and LMWHA samples.

It is well known that the primary function of HA is to hold water and retain a balance of moisture [33]. The critical concern of using LMWHA for wound dressing is the water-absorbing ability, which is reduced when the polymer is depolymerized to small MW fragments [23,24]. In this study, we found that the γ-irradiation reduced the water-absorbing ability when the material came in contact with water (Figure 5). However, after 15 h of experimentation, the water-absorbing abilities of all the HA samples dramatically increased and then reached a plateau at 96 h. At this time, the weight of the water-absorbed by HMWHA was almost the same as that of the HA20I sample. This phenomenon may be due to γ-irradiation and would not affect the water-holding related chemical structure. Accordingly, pre-immersion of the γ-irradiated HA in water for a certain time resolved the problem of water-absorbing ability reduction when the LMWHA was considered for use.

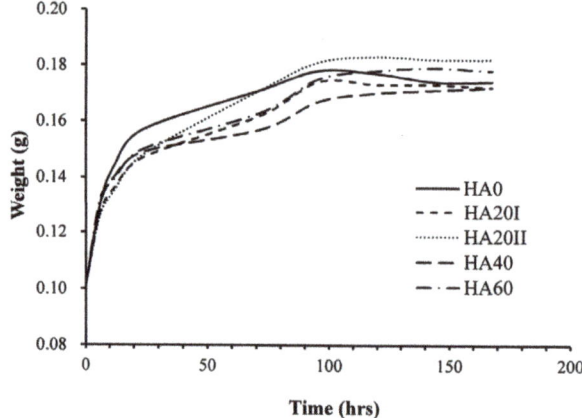

Figure 5. Effect of γ-irradiation on the water-absorbing ability of LMWHA over time.

The cell viability of the HA samples was evaluated using the MTT assay (Figure 6) in L929 fibroblasts. The cells cultured with HMWHA and LMWHA showed typical growth curves. The L929 fibroblasts exhibited no cytotoxicity. During the four-day experimental period, cells cultured with γ-irradiated HA exhibited significantly higher viability compared to the cells cultured with untreated HMWHA ($p < 0.05$). The result that γ-irradiated LMWHA increases skin cell proliferation becomes a crucial property for their application in wound healing [34,35]. Because the γ-irradiated HA exhibited similar properties in the chemical and physical analyses and in vitro cell experimentation, only the HA20I sample was used to fabricate the dressing fabric used in the rat study in keeping with the policy of animal reduction based on the Helsinki Declaration.

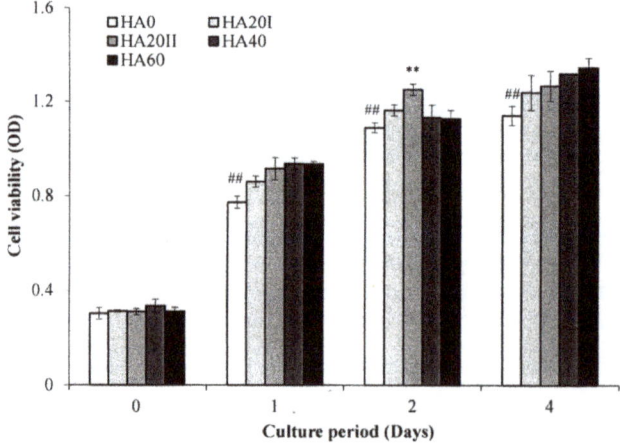

Figure 6. Cell viability tests for the HMWHA and LMWHA irradiated with various dosages of γ-radiation. The data are presented as mean ± SD ($n = 84$). ** and ## denote significant differences ($p < 0.05$).

The CMC/HA fabric was fabricated for wound dressing in this study. Previous studies demonstrated that CMC/HA is nontoxic, nonmutagenic, nonimmunogenic, nonirritating, nonpyrogenic [36] and did not induce an inflammatory cytokine response [37]. In addition, the fabrication of such a CMC/HA hydrogel does not require chemical additives or an energy source [26]. The in vivo wound healing activity of the CMC/LMWHA dressing is shown in Figure 7. Two days after the skin excision, the healing process inside the epidermis was better in the wound covered with CMC/LMWHA fabric compared to the CMC only fabric. The CMC/LMWHA fabric-covered wound showed accelerated healing and lower secretions than in the CMC fabric-only group. The quantitative results demonstrated that the wounds covered with CMC/LMWHA fabric (60.42 ± 5.29%) resulted in a statistically significant reduction ($p < 0.05$) in the wound area when compared to that of the wound covered with CMC fabric (75.07 ± 2.79%) (Figure 8). This statistically significant reduction in wound size was observed on days 4–8 of the experimental period. On day 8, compared to the control group, the wounds covered with CMC/LMWHA fabric became dark brown, dry and smaller than the wounds covered with the CMC fabric (Figure 7). The wound size of the CMC/LMWHA fabric group decreased significantly to 19.23 ± 1.45% (Figure 8), which is almost half that of the wounds covered with CMC fabric (33.12 ± 6.54%) ($p < 0.05$). It is known that antimicrobial activity is an essential requirement for evaluating a wound dressing material [3]. Since the growth of wound-related bacteria requires a neutral pH environment, an acidic environment is not conducive to the growth of bacteria [38,39]. Thus, the reduction in pH value due to HA depolymerization (Table 1) results in an acidic environment, which may be the reason leading to the promotion of skin cell proliferation [40] and collagen reorganization, with the resulting acceleration of wound healing [38,41] as showed in Figures 7 and 8. According to these results, we confirmed that LMWHA promotes wound healing and acts as an excellent wound dressing material for medical application.

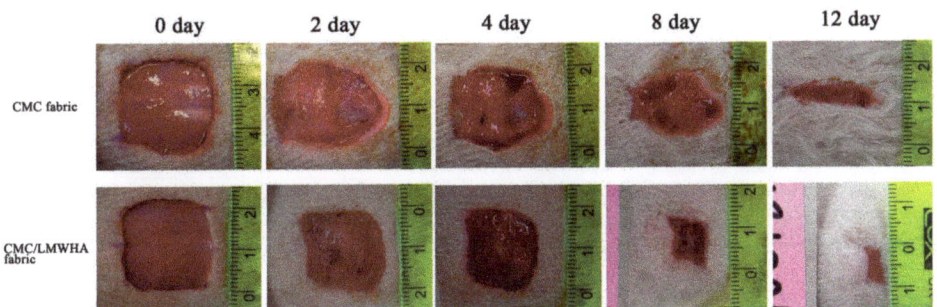

Figure 7. Photographs of the wound in rats after skin excision on days 0, 2, 4, 8 and 12.

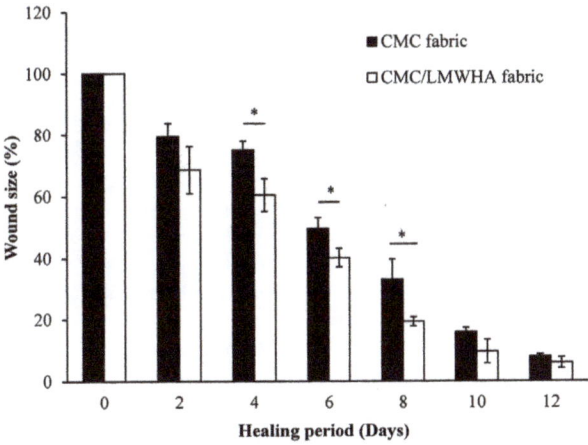

Figure 8. Photographs of the wound in rats after skin excision on days 0, 2, 4, 8 and 12.

4. Conclusions

The LMWHA prepared by γ-irradiation shows strain rate-dependent Newton/non-Newton fluid biphasic viscosity. The water absorption ability of γ-ray-prepared LMWHA is reserved when used for long-term medical application. In addition, the LMWHA prepared by γ-irradiation can be an excellent wound dressing material for medical applications.

Author Contributions: Conceptualization, H.-M.H.; Data curation, Y.-C.H., K.-Y.H. and W.-Z.L.; Funding acquisition, K.-H.F.; Investigation, W.-J.C., Y.-C.H.; Methodology, S.-W.F.; Project administration, H.-M.H.; Writing, H.-M.H.

Funding: This research received no external funding.

Conflicts of Interest: The authors declare no conflict of interest.

References

1. Monsuur, H.N.; Boink, M.A.; Weijers, E.M.; Roffel, S.; Breetveld, M.; Gefen, A.; van den Broek, L.J.; Gibbs, S. Methods to study differences in cell mobility during skin wound healing in vitro. *J. Biomech.* **2016**, *49*, 1381–1387. [CrossRef] [PubMed]
2. Lin, T.Z.; Zhong, L.; Santiago, J.L. Anti-inflammatory and skin barrier repair effects of topical application of some plant oils. *Int. J. Mol. Sci.* **2018**, *19*, 70. [CrossRef] [PubMed]
3. Wang, Y.F.; Que, H.F.; Wang, Y.J.; Cui, X.J. Chinese herbal medicines for treating skin and soft-tissue infections. *Cochrane Database Syst. Rev.* **2014**, *25*, CD010619. [CrossRef] [PubMed]
4. Korn, P.; Schulz, M.C.; Hintze, V.; Range, U.; Mai, R.; Eckelt, U.; Schnabelrauch, M.; Möller, S.; Becher, J.; Scharnweber, D.; et al. Chondroitin sulfate and sulfated hyaluronan-containing collagen coatings of titanium implants influence peri-implant bone formation in a minipig model. *J. Biomed. Mater. Res. A* **2014**, *102*, 2334–2344. [CrossRef] [PubMed]
5. Correia, C.R.; Moreira-Teixeira, L.S.; Moroni, L.; Reis, R.L.; van Blitterswijk, C.A.; Karperien, M.; Mano, J.F. Chitosan scaffolds containing hyaluronic acid for cartilage tissue engineering. *Tissue Eng. Part C Methods* **2011**, *17*, 717–730. [CrossRef] [PubMed]
6. Dahiya, P.; Kamal, R. Hyaluronic acid: A boon in periodontal therapy. *N. Am. J. Med. Sci.* **2013**, *5*, 309–315. [CrossRef]
7. Zhang, W.; Mu, H.; Zhang, A.; Cui, G.; Chen, H.; Duan, J.; Wang, S. A decrease in moisture absorption–retention capacity of N-deacetylation of hyaluronic acid. *Glycoconj. J.* **2013**, *30*, 577–583. [CrossRef]
8. Kablik, J.; Monheit, G.D.; Yu, L.; Chang, G.; Gershkovich, J. Comparative physical properties of hyaluronic acid dermal fillers. *Dermatol. Surg.* **2009**, *35* (Suppl. S1), 302–312. [CrossRef]
9. Jentsch, H.; Pomowski, R.; Kundt, G.; Göcke, R. Treatment of gingivitis with hyaluronan. *J. Clin. Periodontol.* **2003**, *30*, 159–164. [CrossRef]
10. Frenkel, J.S. The role of hyaluronan in wound healing. *Int. Wound J.* **2014**, *11*, 159–163. [CrossRef]
11. D'Agostino, A.; Stellavato, A.; Busico, T.; Papa, T.; Tirino, V.; Papaccio, G.; La Gatta, A.; De Rosa, M.; Schiraldi, C. In vitro analysis of the effects on wound healing of high- and low-molecular weight chains of hyaluronan and their hybrid H-HA/L-HA complexes. *BMC Cell Biol.* **2015**, *16*, 19. [CrossRef] [PubMed]
12. Voigt, J.; Driver, V.R. Hyaluronic acid derivatives and their healing effect on burns, epithelial surgical wounds, and chronic wounds: A systematic review and meta-analysis of randomized controlled trials. *Wound Repair Regen.* **2012**, *20*, 317–331. [CrossRef] [PubMed]
13. Neuman, M.G.; Nanau, R.M.; Oruña-Sanchez, L.; Coto, G. Hyaluronic acid and wound healing. *J. Pharm. Pharm. Sci.* **2015**, *18*, 53–60. [CrossRef] [PubMed]
14. Cowman, M.K.; Hittner, D.M.; Feder-Davis, J. ^{13}C-NMR studies of hyaluronan: Conformational sensitivity to varied environments. *Macromolecules* **1996**, *29*, 2894–2902. [CrossRef]
15. Ke, C.; Sun, L.; Qiao, D.; Wang, D.; Zeng, X. Antioxidant activity of low molecular weight hyaluronic acid. *Food Chem. Toxicol.* **2011**, *49*, 2670–2675. [CrossRef] [PubMed]
16. Maharjan, A.S.; Pilling, D.; Gomer, R.H. High and low molecular weight hyaluronic acid differentially regulate human fibrocyte differentiation. *PLoS ONE* **2011**, *6*, e26078. [CrossRef] [PubMed]

17. Rayahin, J.E.; Buhrman, R.S.; Zhang, Y.; Koh, T.J.; Gemeinhart, R.A. High and low molecular weight hyaluronic acid differentially influence macrophage activation. *ACS Biomater. Sci. Eng.* **2015**, *13*, 481–493. [CrossRef]
18. Kavasi, R.M.; Berdiaki, A.; Spyridaki, I.; Corsini, E.; Tsatsakis, A.; Tzanakakis, G.; Nikitovic, D. HA metabolism in skin homeostasis and inflammatory disease. *Food Chem. Toxicol.* **2017**, *101*, 128–138. [CrossRef] [PubMed]
19. Trabucchi, E.; Pallotta, S.; Morini, M.; Corsi, F.; Franceschini, R.; Casiraghi, A.; Pravettoni, A.; Foschi, D.; Minghetti, P. Low molecular weight hyaluronic acid prevents oxygen free radical damage to granulation tissue during wound healing. *Int. J. Tissue React.* **2002**, *24*, 65–71.
20. Gariboldi, S.; Palazzo, M.; Zanobbio, L.; Selleri, S.; Sommariva, M.; Sfondrini, L.; Cavicchini, S.; Balsari, A.; Rumio, C. Low molecular weight hyaluronic acid increases the self-defense of skin epithelium by induction of-defensin 2 via TLR2 and TLR4. *J. Immunol.* **2008**, *181*, 2103–2110. [CrossRef]
21. Chen, H.; Qin, J.; Hu, Y. Efficient degradation of high-molecular-weight hyaluronic acid by a combination of ultrasound, hydrogen peroxide, and copper ion. *Molecules* **2019**, *24*, 617. [CrossRef] [PubMed]
22. Hokputsa, S.; Jumel, K.; Alexander, C.; Harding, S.E. A comparison of molecular mass determination of hyaluronic acid using SEC/MALLS and sedimentation equilibrium. *Eur. Biophys. J.* **2003**, *32*, 450–456. [CrossRef] [PubMed]
23. Choi, J.; Kim, J.K.; Kim, J.H.; Kweon, D.K.; Lee, J.W. Degradation of hyaluronic acid powder by electron beam irradiation, gamma ray irradiation, microwave irradiation and thermal treatment: A comparative study. *Carbohydr. Polym.* **2010**, *79*, 1080–1085. [CrossRef]
24. Kim, J.K.; Sung, N.Y.; Srinivasan, P.; Choi, J.I.; Kim, S.K.; Oh, J.M.; Kim, J.H.; Song, B.S.; Park, H.J.; Byun, M.W.; et al. Effect of gamma irradiated hyaluronic acid on acetaminophen induced acute hepatotoxicity. *Chem. Biol. Interact.* **2008**, *172*, 141–153. [CrossRef] [PubMed]
25. Yue, W. Preparation of low-molecular-weight hyaluronic acid by ozone treatment. *Carbohydr. Polym.* **2012**, *89*, 709–712.
26. Huang, Y.C.; Huang, K.U.; Yang, B.Y.; Ko, C.H.; Huang, H.M. Fabrication of novel hydrogel with berberine-enriched carboxymethylcellulose and hyaluronic acid as an anti-inflammatory barrier membrane. *BioMed Res. Int.* **2016**, 3640182. [CrossRef] [PubMed]
27. Scott, J.E.; Heatley, F. Hyaluronan forms specific stable tertiary structures in aqueous solution: A 13C NMR study. *Proc. Natl. Acad. Sci. USA* **1999**, *96*, 4850–4855. [CrossRef]
28. Scott, J.E.; Heatley, F. Biological properties of hyaluronan in aqueous solution are controlled and sequestered by reversible tertiary structures, defined by NMR Spectroscopy. *Biomacromolecules* **2002**, *3*, 547–553. [CrossRef]
29. Nagasawa, N.; Mitomo, M.; Yoshii, F.; Kume, T. Radiation-induced degradation of sodium alginate. *Polym. Degrad. Stab.* **2000**, *69*, 279–285. [CrossRef]
30. Gura, E.; Huckel, M.; Muller, P.J. Specific degradation of hyaluronic acid and its rheological properties. *Polym. Degrad. Stab.* **1998**, *59*, 297–302. [CrossRef]
31. Lapčík, L., Jr.; Benešová, K.; Lapčík, L.; De Smedt, S.; Lapčíková, B. Chemical modification of hyaluronic acid: Alkylation. *Int. J. Polym. Anal. Charact.* **2010**, *15*, 486–496. [CrossRef]
32. Haward, S.J.; Jaishankar, A.; Oliveira, M.S.N.; Alves, M.A.; McKinley, G.H. Extensional flow of hyaluronic acid solutions in an optimized microfluidic cross-slot device. *Biomicrofluidics* **2013**, *7*, 044108. [CrossRef] [PubMed]
33. Necas, J.; Bartosikova, L.; Brauner, P.; Kolar, J. Hyaluronic acid (hyaluronan): A review. *Vet. Med. Czech* **2008**, *53*, 397–411. [CrossRef]
34. Lönnroth, E.C. Toxicity of medical glove materials: A pilot study. *Int. J. Occup. Saf. Ergon.* **2005**, *11*, 131–139. [CrossRef] [PubMed]
35. Eskandarinia, A.; Kefayat, A.; Rafienia, M.; Agheb, M.; Navid, S.; Ebrahimpour, K. Cornstarch-based wound dressing incorporated with hyaluronic acid and propolis: In vitro and in vivo studies. *Carbohydr. Polym.* **2019**, *15*, 25–35. [CrossRef]
36. Burns, J.W.; Colt, M.J.; Burgees, L.S.; Skinner, K.C. Preclinical evaluation of Seprafilm bioresorbable membrane. *Eur. J. Surg. Suppl.* **1997**, *577*, 40–48.
37. Uchida, K.; Otake, M.; Inoue, Y.; Koike, K.; Matsushita, K.; Tanaka, K.; Inoue, Y.; Mohri, Y.; Kusunoki, M. Bacteriostatic effects of hyaluronan-based bioresorbable membrane. *Surg. Sci.* **2001**, *2*, 431–436. [CrossRef]

38. Rekik, D.M.; Khedir, S.B.; Moalla, K.K.; Kammoun, N.G.; Rebai, T.; Sahnoun, Z. Evaluation of wound healing properties of grape seed, sesame, and fenugreek oils. *Evid. Based Complement. Altern Med.* **2016**, *2016*, 7965689.
39. Saeidnia, S.; Manayi, A.; Gohari, A.R.; Abdollahi, M. The story of beta-sitosterol—A review. *Eur. J. Med. Plants* **2014**, *4*, 590–609. [CrossRef]
40. Chen, C.C.; Nien, C.J.; Chen, L.G.; Huang, K.Y.; Chang, W.J.; Huang, H.M. Effects of Sapindus mukorossi seed oil on skin wound healing: In vivo and in vitro testing. *Int. J. Mol. Sci.* **2019**, *20*, 2579. [CrossRef]
41. Mwipatayi, B.P.; Angel, D.; Norrish, J.; Hamilton, M.J.; Scott, A.; Sieunarine, K. The use of honey in chronic leg ulcers: A literature review. *Prim. Intent.* **2004**, *12*, 107–112.

© 2019 by the authors. Licensee MDPI, Basel, Switzerland. This article is an open access article distributed under the terms and conditions of the Creative Commons Attribution (CC BY) license (http://creativecommons.org/licenses/by/4.0/).

MDPI
St. Alban-Anlage 66
4052 Basel
Switzerland
Tel. +41 61 683 77 34
Fax +41 61 302 89 18
www.mdpi.com

Polymers Editorial Office
E-mail: polymers@mdpi.com
www.mdpi.com/journal/polymers

www.ingramcontent.com/pod-product-compliance
Lightning Source LLC
LaVergne TN
LVHW070659100526
838202LV00013B/1000